# Synthesis Lectures on Wave Phenomena in the Physical Sciences

**Series Editor**

Sanichiro Yoshida, Department of Chemistry and Physics, Southeastern Louisiana University, Hammond, LA, USA

The aim of this series is to discuss the science of various waves. An emphasis is laid on grasping the big picture of each subject without dealing formalism, and yet understanding the practical aspects of the subject. To this end, mathematical formulations are simplified as much as possible and applications to cutting edge research are included.

Sanichiro Yoshida

# Physics and Mathematics Behind Wave Dynamics

 Springer

Sanichiro Yoshida
Department of Chemistry and Physics
Southeastern Louisiana University
Hammond, LA, USA

ISSN 2690-2346          ISSN 2690-2354  (electronic)
Synthesis Lectures on Wave Phenomena in the Physical Sciences
ISBN 978-3-031-60356-3          ISBN 978-3-031-60354-9  (eBook)
https://doi.org/10.1007/978-3-031-60354-9

*This book is dedicated to the memory of my late father, Akira Yoshida, who introduced me to the joy of science and engineering; my mother, Sonoko Yoshida, who taught me how to learn; and my wife, Yuko Yoshida, for filling my life with love, joy, and happiness.*

# Preface

This book grew from my laboratory notebook. I conducted various experimental and theoretical research projects using a wave to characterize the properties of physical systems, e.g., ultrasonic nondestructive testing to probe anomalies in metal materials, an optical interferometric research project to diagnose the deformation status of solid materials, a development of signal processing algorithms to improve the performance of cochlear implant devices, and gravitational wave detection with an optical interferometer.

Through these activities, waves and wave dynamics have fascinated me. At the same time, I have encountered difficulties in treating the wave signals. Often, the signal-to-noise ratio is beyond my control. Another time, the signal is too complicated to process. These experiences motivated me to write this book.

The reader may notice that half of this book discusses frequency domain analysis, such as Fourier Transform, Laplace Transform, and Transfer Functions. I have a reason for this. During these times, I have learned frequency domain analysis can ease these difficulties. By dividing the signal into different frequency ranges, we can interpret the observed signal better and have better insight into the underlying phenomenon.

The organization of this book is as follows. Chapter 1 introduces waves as a moving oscillatory pattern. Chapter 2 focuses on the concept of harmonic oscillation. In Chap. 3, we discuss the mathematical aspect of wave dynamics. We derive wave equations and solve them. Chapter 4 discusses some basic properties of waves. In Chaps. 5 and 6, we view waves in the frequency domain. We discuss various mathematical concepts related to frequency domain analysis.

I hope this book helps students to deepen their understanding of wave dynamics and apply the knowledge to their research activities.

Hammond, LA, USA                                                  Sanichiro Yoshida
January 2024

# Contents

We can view a wave as a snapshot of a periodic pattern moving in a certain direction at a constant velocity. We can also view a wave as a pattern in which different locations experience the same temporal oscillation with a certain time lag. Take a wave formed by the spectators in a soccer stadium as an example. If you look at the wave from a helicopter hovering above the stadium, you will see a periodic pattern of people's hands going around the stadium. This is the first view. If you look at several spectators next to each other, you will see each of them move their hands up and down at the same rate with a certain delay. When the first person in the group raises his or her hands at the highest position, the hands of the person on the left are about to reach the highest position, reaching the highest position a few seconds later. This is the second view of the wave. In this case, the wave travels toward the right of this group of spectators.

In this chapter, we discuss waves according to these two views. We will find that the spatial and temporal periodic patterns constitute a wave motion. Several basic quantities associated with these views of a wave will be introduced. Some mathematical tools that facilitate the description of waves are also discussed.

## 1.1 Wave as a Moving Oscillatory Pattern

We can view a wave as an oscillation moving in space. The spectator of a soccer stadium can make a wave because one person's up-and-down hand motion moves around the stadium. If all the spectators move their hands simultaneously, no wave is formed. It is simply an oscillation.

© The Author(s), under exclusive license to Springer Nature Switzerland AG 2025

S. Yoshida, *Physics and Mathematics Behind Wave Dynamics*, Synthesis Lectures on Wave Phenomena in the Physical Sciences,

https://doi.org/10.1007/978-3-031-60354-9_1

Mathematically, we can express the above fact as follows. Suppose $f(t, x)$ is a periodic function of time $t$ and space $x$ where the period is $\theta$.

$$f(\phi) = f(\phi \pm n\theta) \tag{1.1}$$

Here $\phi$ is the phase and $n$ is an integer. The left-hand side of (1.1) represents the value of function $f(t, x)$ when the combination of time and position makes the phase equal to $\phi$. The entire Eq. (1.1) expresses that the value of the function is the same when the phase varies by an integer multiple of period $\theta$. Here $\theta$ is a constant.

Consider that $\phi$ is a linear function of the time and spatial coordinates as

$$\phi = \omega t \pm k x \tag{1.2}$$

where $\omega$ and $k$ are constants whose meanings will be clarified shortly. At a fixed location $x$, the phase varies as a function of $t$. Selecting this location to be $x = 0$ does not lose the generality. Under this condition, from (1.1) and (1.2) we find it as follows.

$$f(\omega t) = f(\omega t + \theta) = f(\omega(t + \tau)) \tag{1.3}$$

Expression (1.3) indicates that every $\tau$ s (seconds) the function $f(t, 0)$ takes the same value. Thus, we can interpret $\tau$ to be the period in time. (We can say that $\theta$ is the period in phase.) The same argument holds at an arbitrary location $x_0$,

$$f(\omega t + k x_0) = f(\omega t + k x_0 + \theta) = f(\omega(t + \tau) + k x_0) \tag{1.4}$$

At the fixed location $x_0$, the phase $\phi$ varies only depending on $t$. So, the function $f(\phi)$ takes the same value every $\tau$ s.

Since $\phi$ is a linear function of $t$, $\omega(t + \tau) = \omega t + \omega \tau$. So, from (1.3) we find the following equation.

$$\tau = \frac{\theta}{\omega} \tag{1.5}$$

The reciprocal of $\tau$ indicates how many times the oscillation occurs in 1 s, i.e., how frequent the oscillation is.

$$\nu \equiv \frac{1}{\tau} \tag{1.6}$$

This quantity $\nu$ is referred to as the frequency of oscillation and its SI unit (Système International d'unités, or the International System of Units) is Hz (= 1/s). The quantity $\omega$ is called the angular frequency (sometimes called the frequency as well) and its unit is rad/s.

Repeating the same argument as above, we can define the spatial periodicity $\lambda$ as follows.

$$\lambda = \frac{\theta}{k} \tag{1.7}$$

The quantity $\lambda$ is referred to as the wavelength. It is the spatial period of the wave in the SI unit of m. The quantity $k$ is called the wave number. It is the spatial frequency of the wave in the unit of rad (radian).

## 1.2    Phase Velocity

Multiplication of (1.6) and (1.7) yields the following equation.

$$
v\lambda = \frac{\omega}{k} \tag{1.8}
$$

The left-hand side of (1.8) indicates how many times the wavelength repeats in 1 s. If the observer stays at a fixed place, he will see the pattern of the spatial periodicity passing him $v\lambda$ times a second. Changing the perspective, we can also say that (1.8) represents the velocity of an observer that sits on the wave at a constant phase. For this observer, the phase does not change. Therefore, $d\phi/dt = 0$. By differentiating $\phi$ with respect to time, we obtain the following expression.

$$
\frac{d\phi}{dt} = \omega \pm k \frac{dx}{dt} \tag{1.9}
$$

Setting (1.9) to zero, we obtain the following equation.

$$
\frac{dx}{dt} = \pm \frac{\omega}{k} \tag{1.10}
$$

We can interpret that the left-hand side of (1.10) represents the change in the position of a constant phase over time, and call it the phase velocity of the wave. (1.9) indicates that the phase velocity is the product of the frequency and wavelength of a wave.

$$
v_p = v\lambda = \pm \frac{\omega}{k} \tag{1.11}
$$

The sign in front of $\omega/k$ on the right-hand side of (1.11) represents the direction of the wave's motion.

By substituting (1.2) into the expression on the left-hand side of (1.1) and using (1.11), we can express the wave function $f$ in the following way.

$$
f(\phi) = f(\omega t \pm kx) = f\left(\omega\left(t - \frac{x}{\pm v_p}\right)\right) = f\left(k\left(\pm v_p t - x\right)\right) \tag{1.12}
$$

Consider the constant phase condition, $\phi = \phi_0$, in (1.12).

$$
\omega t \pm kx = \phi_0 \tag{1.13}
$$

If the sign in front of $kx$ is positive, $x$ must decrease to keep the phase constant at $\phi_0$ over time (when $t$ increases). This means that the wave moves in the negative direction along the

$x$-axis. If it is negative, the wave moves in the positive $x$ direction. Note that the former case where the wave moves in the negative $x$ direction corresponds to $v_p < 0$, and the latter case where the wave moves in the positive $x$ direction corresponds to $v_p > 0$ on the right-hand side of (1.12). Thus, allowing for the phase velocity to take a positive or negative value, we can express the wave function in the following form.

$$f(\phi) = f\left(\omega\left(t - \frac{x}{v_p}\right)\right) = f\left(k\left(v_p t - x\right)\right) \tag{1.14}$$

In the form of (1.14), a positive $v_p$ represents a wave moving in the positive $x$ direction and a negative $v_p$ represents a wave in the negative $x$ direction.

## 1.3    Amplitude and Phase

The concepts of the amplitude and phase of an oscillatory system are most naturally and easily understood in association with a circular motion. Consider in Fig. 1.1 that point $P$ moves along the circumference of the circle of radius $A$ with a constant angular velocity of $\omega$. The angle subtended by the arc between points $P$ and $(x, y) = (A, 0)$ varies linearly with time as

$$\theta(t) = \omega t + \phi_0 \tag{1.15}$$

Here $\phi_0$ is the initial angle at $t = 0$. The projection of point $P$ onto the $x$-axis varies as

$$P_x(t) = A \cos\theta(t) = A \cos(\omega t + \phi_0) \tag{1.16}$$

In the next chapter, we will discuss the harmonic oscillation of a point mass connected to a spring. There we describe the dynamics using the displacement of the mass from its equilibrium position, $\xi(t)$. By viewing the origin $(x, y) = (0, 0)$ as the equilibrium point and $P_x(t)$ as the position of the mass connected to the spring as discussed above, we can

**Fig. 1.1** Point $P$ moves along the circumference at a constant circular speed

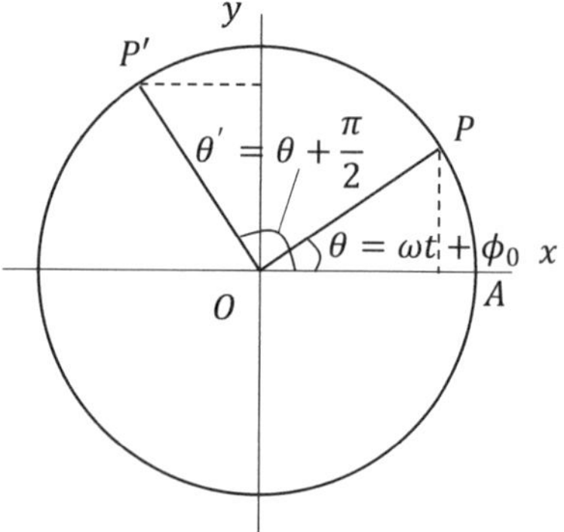

relate $P_x(t)$ to the displacement $\xi(t)$ in (2.4). In this context, the phase $\omega t + \phi_0$ is the angle subtended by the arc between the current and initial positions of $P$ on the circumference. Over time by $dt$, $\omega t$ part of the phase $\theta(t)$ increases by $\omega dt$. Accordingly, the point $P$ moves counterclockwise on the circumference. Thus, we can relate the passage of time to the counterclockwise rotation of the radius $OP$, where the $x$ projection represents the time-varying displacement of the mass $m$ from the equilibrium.

We can repeat the same argument by considering the projection to the $y$-axis. In this case, the displacement from the origin along the $y$-axis becomes

$$P_y(t) = A \sin \theta(t) \tag{1.17}$$

Thus, a harmonic oscillation can be interpreted as the projection of a point under constant circular motion onto a coordinate axis. If we use the angle from the positive $y$-axis, $\theta'$, instead of the angle from the positive $x$-axis, expression (1.17) becomes as follows.

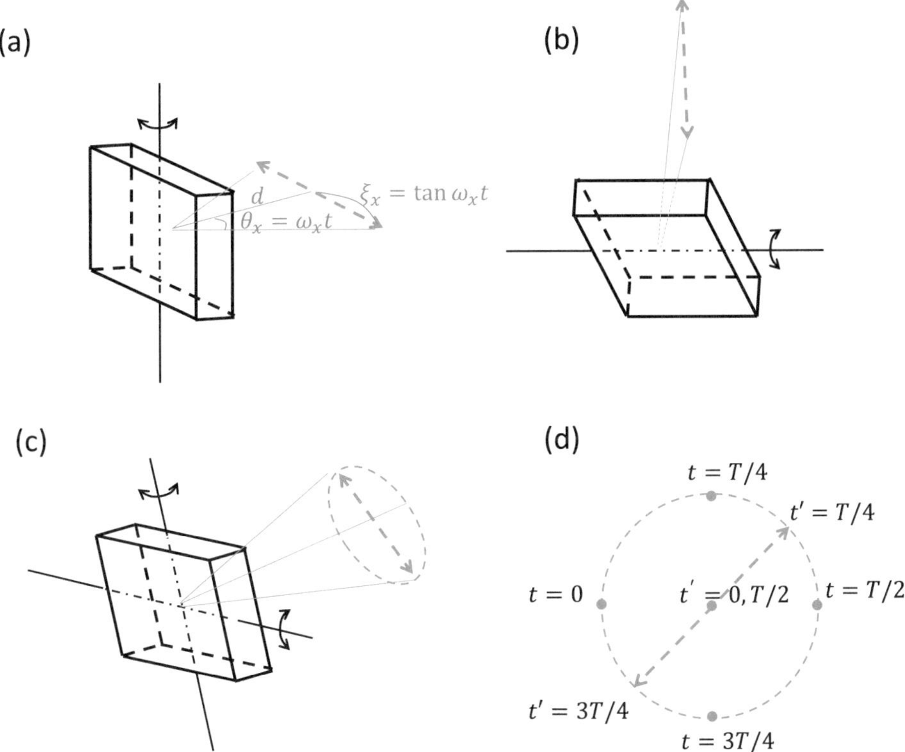


Fig. 1.2 Laser beam reflected by a mirror experiencing **a** yaw motion, **b** pitch motion, and **c** yaw and pitch motion. **d** the trajectory of the reflected laser beam on the screen. $t = 0, T/4, T/2, 3T/4$ represent the spot at four representative moments when yaw and pitch motion have a phase difference of $\pi/2$. Here $T$ is period. $t'$ represents the times when yaw and pitch are in phase

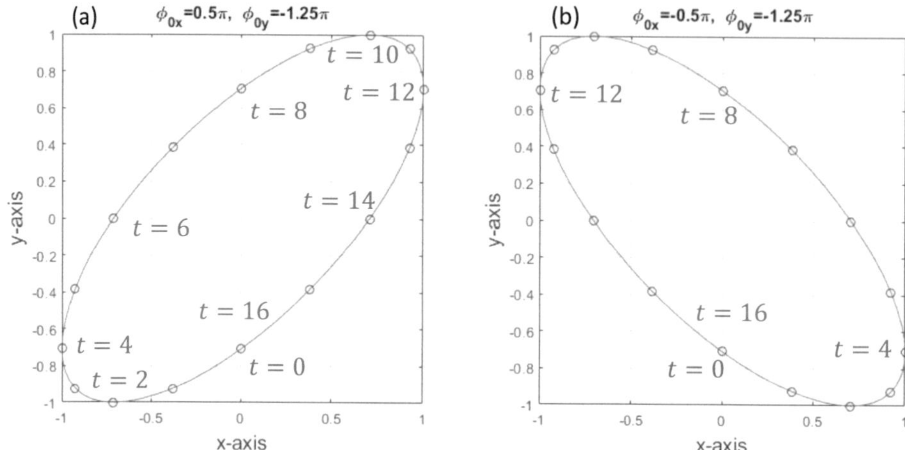


**Fig. 1.3** Trajectory of the tip of a vector whose $x$ and $y$ components undergo oscillation of the same frequency with a phase difference

$$P_y(t) = A \cos \theta'(t) = A \cos \left( \omega t - \frac{\pi}{2} \right) \tag{1.18}$$

Here, $\pi/2 = \theta - \theta'$ can be interpreted as the phase difference between the harmonic oscillations associated with the $x$ projection and $y$ projection of the same circular motion. This is a geometric presentation of the phase shift discussed above.

Above, we argued the horizontal and vertical harmonic oscillations as the $x$ and $y$ projections of the same circular motion. Each of the two oscillations may be due to mutually independent harmonic oscillation. In this case, the phase delay $\phi_0$ can be any value. Moreover, the displacement vector associated with each oscillation can be oriented in any direction. It is particularly important to consider the case where these two displacement vectors constitute the $x$ and $y$ components of the displacement vector representing the same harmonic oscillation; the $x$ and $y$ components of the harmonic oscillation have a phase difference. Since the two oscillations belong to the same harmonic oscillation, they have the same angular frequency $\omega$. An example is the situation where a mirror undergoing pitch and yaw motions with a phase difference and you monitor the reflection of a laser beam off the mirror surface on a screen. Figure 1.2 illustrates such a setup schematically.

Consider a sample situation in Fig. 1.3 where the $x$ and $y$ components of the vector undergo oscillation of the same angular frequency with a phase difference of $\pi/4$. Note that the horizontal and vertical axes of this figure represent the geometrical orientation, not the phase delay of $\pi/2$ as is the case of Fig. 1.1. The trajectory of the tip of the total vector $\boldsymbol{\xi} = \xi_x \hat{x} + \xi_y \hat{y}$ is an ellipse. This concept becomes important when we discuss birefringence in a later chapter.

## 1.4 Euler's Formula (Notation)

What we discussed so far in this chapter can be conveniently expressed with the use of complex numbers. The following expression known as Euler's formula (notation) is the base of the concept [1].

$$Ae^{i\theta} = A\left(\cos\theta + i\sin\theta\right) \tag{1.19}$$

Here $i$ is the imaginary unit defined as $i^2 = -1$. Call the complex number expressed by (1.19) $c$ and consider it on a complex plane in Fig. 1.4.

From the right-hand side of (1.19), the real and imaginary parts of $c$ are

$$Re(c) = A\cos\theta \tag{1.20}$$

$$Im(c) = A\sin\theta \tag{1.21}$$

The absolute value of this complex number is, by definition, as follows.

$$|c| = \sqrt{A^2\cos^2\theta + A^2\sin^2\theta} = A \tag{1.22}$$

So, the amplitude of the harmonic oscillation can be identified as the absolute value of the complex representation. The phase $\theta$, on the other hand, is the angle of $c$ from the positive real axis.

$$\theta = \tan^{-1}\frac{Im(c)}{Re(c)} \tag{1.23}$$

The use of the complex form greatly simplifies the treatment of the phase change as a function of time. Substitute (1.15) into (1.19).

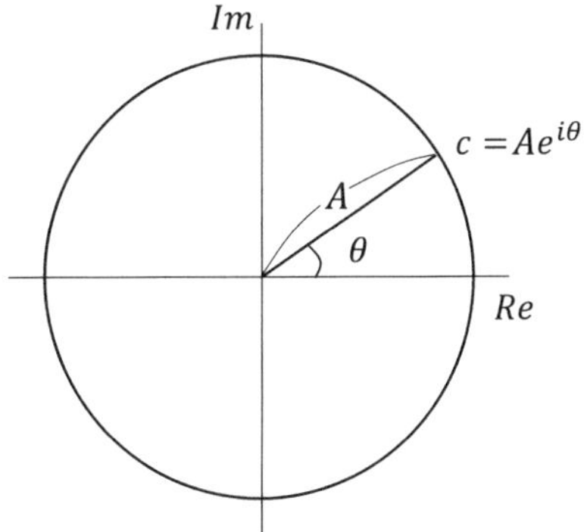


**Fig. 1.4** Complex number $c$ on a complex plane

$$Ae^{i(\omega t+\phi_0)} = A\left(\cos(\omega t + \phi_0) + i\sin(\omega t + \phi_0)\right) \tag{1.24}$$

Note that in the complex form, the real part is physically meaningful. In this form, the passage of time $dt$ corresponds to the change in the left-hand side as follows.

$$Ae^{i(\omega(t+dt)+\phi_0)} = Ae^{i(\omega t+\phi_0)}e^{i\omega dt} \tag{1.25}$$

Equation (1.25) indicates that the passage of time $dt$ can be expressed as the multiplication of phase $\delta = \omega dt$.

$$e^{i\omega dt} \equiv e^{i\delta}. \tag{1.26}$$

Here $e^{i\delta}$ is referred to as the phase factor. The phase factor changes the real part of the complex number, and the change is sinusoidal.

Temporal differentiation and integration can also be handled easily with the complex notation. Differentiate both-hand sides of (1.24). For simplicity, put the initial phase $\phi_0 = 0$.

$$Ae^{i\omega t}(i\omega) = A\omega\left(-\sin\omega t + i\cos\omega t\right), \tag{1.27}$$

The left-hand side of (1.27) indicates that temporal differentiation can be expressed by the multiplication of $i\omega$ if we use the complex form. Putting $\phi_0 = 0$ means that the initial position of the mass is at the turning point on the positive $x$ side. The right-hand side indicates that the velocity of the mass is in the negative $x$ direction until the mass comes to the other turning point and that the velocity is positive from that point until the mass comes back to the first turning point. It also indicates that if the displacement is cosine-like the velocity is sine-like.

According to (1.19), the multiplication of $i = e^{i\pi/2}$ is equivalent to advancing the phase by $\pi/2$. Thus, the temporal differentiation is equivalent to advancing the phase by $\pi/2$ and scaling the magnitude by the factor of $\omega$.

Similarly, temporal integration can be argued as follows.

$$\frac{Ae^{i\omega t}}{(i\omega)} = \frac{A}{\omega}\left(\sin\omega t - i\cos\omega t\right), \tag{1.28}$$

Now since $1/i = -i = e^{-i\pi/2}$, we find that the integration is equivalent to retarding the phase by $\pi/2$.

Thus, Euler's notation makes our life much easier in the context of solving equations of motion, as we will discuss in Chap. 2.

## Reference

1. Boas ML (2006) Mathematical methods in the physical sciences, 3rd edn. Wiley, London, p 61

# Harmonic Oscillations 2

In an oscillatory system, the oscillating entity (called the particle) passes the equilibrium (neutral) position and changes its direction at the turning points. The harmonic oscillation is one of the most fundamental forms of such an oscillatory motion. A sine or cosine function whose argument is a linear function of time generally expresses a harmonic oscillation. Fourier theorem states that a real function can be expanded as a series of sine, cosine, and their harmonics [1–3]. Therefore, the analysis of harmonic oscillation is imperative to understand oscillation dynamics.

In this chapter, we discuss harmonic oscillations. Using a simple spring-mass system as an example, we discuss the harmonic oscillation of the point-mass in detail. The equation of motion that governs the dynamics of the point mass is derived under various conditions and solved in several fashions. Other typical examples of harmonic oscillations are also presented.

## 2.1 Harmonic Oscillation in a Few Words

The harmonic oscillation can be characterized by the following conditions.
(a) The magnitude of the velocity decreases as the particle passes the neutral (equilibrium) point and approaches zero at the turning points.
(b) The sign of the acceleration changes every time the particle passes the neutral (equilibrium) point.

Alternatively, we can express (a) and (b) as follows.
(c) The magnitude of the velocity is at the maximum when the displacement is zero and the magnitude is zero at the turning points where the acceleration is at the maximum as the velocity changes its sign.

© The Author(s), under exclusive license to Springer Nature Switzerland AG 2025

S. Yoshida, *Physics and Mathematics Behind Wave Dynamics*, Synthesis Lectures on Wave Phenomena in the Physical Sciences,

https://doi.org/10.1007/978-3-031-60354-9_2

(d) The sign of the acceleration is opposite to that of the position (or the displacement from the equilibrium point).

The sine and cosine functions beautifully satisfy these conditions. For example, consider that the displacement from the equilibrium is a sine-like function of time. Its temporal derivative, the cosine function, is zero when the sine function takes either the maximum or minimum value. Conversely, the cosine function takes the maximum or minimum when the sine function is zero. Remembering that the temporal derivative of a displacement function is velocity, we see the property (c) in these statements. The second-order temporal derivative of the displacement is negative sine. Here, we see that the sine and cosine functions beautifully have the pattern that the acceleration and displacement have mutually opposite signs, or the condition (d) mentioned above.

## 2.2　Spring-Mass Systems

In dynamics, the force that embodies the above conditions (a)–(d) is known as elastic force. A spring-point-mass system is the simplest yet informative system to describe the motion of a point mass under elastic force. Figure 2.1 illustrates such a system where a point mass is connected to a single spring and undergoes a harmonic oscillation. Let $m$ be the mass of the point-mass, $\xi$ be the displacement of the point-mass from the equilibrium, and $f_{tot}$ be the total external force on $m$. We can express the equation of motion for this system as follows.

$$m\frac{d^2\xi(t)}{dt^2} = f_{tot} \tag{2.1}$$

Since the spring force is the only external force acting on the point mass, we obtain the following expression for $f_{tot}$.

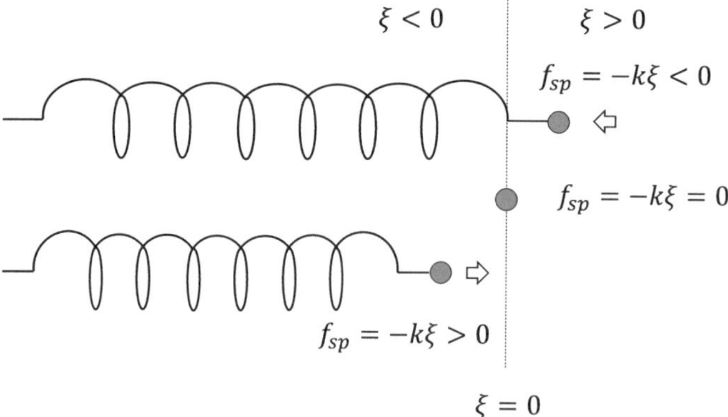


**Fig. 2.1** Spring-mass system consisting of a simple spring and a point mass

$$f_{tot} = f_{sp} = -k\xi(t) \tag{2.2}$$

Here $k$ is thespring constant and $f_{sp}$ represents the elastic force exerted by the spring. Substituting the total external force expression (2.2) into (2.1), we obtain the equation of motion as follows.

$$m\frac{d^2\xi(t)}{dt^2} = -k\xi(t) \tag{2.3}$$

In the following sections, we solve this equation of motion and discuss the properties of the solution.

### 2.2.1  Intuitive Solution

Above, we discussed that the sine and cosine functions possess the characteristics to represent a harmonic oscillation. So, it is reasonable to start our discussion by using the following test function as a solution to the equation of motion (2.3).

$$\xi(t) = A\cos(\omega t + \phi_0) \tag{2.4}$$

It is worthwhile considering a property of the cosine or sine function as a solution to the equation of motion (2.3). This differential equation tells us as follows. If you differentiate the displacement function $\xi(t)$ twice, the result is the function multiplied by a negative constant $-k/m$. On the other hand, if we differentiate $\cos(\omega t + \phi_0)$ or $\sin(\omega t + \phi_0)$ twice, the result is $-\omega^2$ times the original cosine or sine function. This argument justifies that a cosine or sine function can be a solution to (2.3). You may imagine that the equality $k/m = \omega^2$ resulting from (2.3) characterizes the spring-mass system. Later in this section, we will find that the quantity $k/m$ represents the natural frequency of the spring-mass system.

Solution (2.4) has three parameters, the amplitude $A$, the oscillation (angular) frequency $\omega$, and the initial phase $\phi_0$. Here, $A$ tells us how far the point mass can go from the equilibrium, $\omega$ indicates how fast the point mass oscillates, and $\phi_0$ specifies the location of the mass at $t = 0$, i.e., $\xi(0)$. Of these three parameters, only $\omega$ represents the nature of the physical system. Both $A$ and $\phi_0$ are determined by how the external agent starts the oscillation. Below, we will discuss these parameters one by one.

#### Amplitude

Since the cosine function takes the maximum and minimum of $+1$ and $-1$, the range of the harmonic oscillation (2.4) is $\pm A$, regardless of the value of $\omega$ or $\phi_0$. The two extreme positions $\xi = A$ and $\xi = -A$ are called the positive and negative turning points of the oscillator.

The value of $A$ is determined by the initial action provided by the external agent that initiates the oscillation. If the agent initiates the oscillation by pulling or pushing the point mass at a certain location and releasing it, the distance from the equilibrium to this point is the amplitude. The greater this distance, the larger the amplitude $A$. If the agent initiates the oscillation by hitting the point mass at the equilibrium, the harder the agent hits, the larger the amplitude. Thus, $A$ is determined independent of the spring constant $k$ or mass $m$, except that the higher the spring constant the more work must be done by the external agent to initiate the oscillation.

**Initial Phase**

This parameter presents the phase of the oscillatory motion expressed by the cosine function at $t = 0$. Thus, it is called the initial phase. Like the case of amplitude, the external agent determines the initial phase. If the agent initiates the motion at the positive turning point, the displacement at $t = 0$ is A, i.e., $\xi(0) = \cos(\phi_0) = A$. Hence, $\phi_0 = 0$. Similarly, if the external agent initiates the oscillation by hitting the point mass when it is at the equilibrium, $\xi(0) = 0$; it follows that $\phi_0 = \pm\pi/2$ as $\cos(0 \pm \pi/2) = 0$.

**Initial Condition**

You may notice that $\xi(0) = 0$ does not describe the initial motion of the point mass precisely. The point mass can move toward either the positive or negative turning point. To clarify this initial direction, we need to consider the velocity of the point mass, $v(t)$. From (2.4), we find the following expression for $v(t)$.

$$v(t) = d\xi/dt = -\omega A \sin(\omega t + \phi_0) \tag{2.5}$$

Now use the initial phase $\phi_0 = \pm\pi/2$ for the $\xi(0) = 0$ condition separately.

$$\phi_0 = -\pi/2, \quad \xi(0) = A\cos(-\pi/2) = 0, \quad v(0) = -\omega A \sin(-\pi/2) = \omega A > 0 \tag{2.6}$$

$$\phi_0 = \pi/2, \quad \xi(0) = A\cos(\pi/2) = 0, \quad v(0) = -\omega A \sin(\pi/2) = \omega A < 0 \tag{2.7}$$

Find that $\phi_0 = -\pi/2$ represents that the initial velocity is positive $v(0) > 0$ whereas $\phi_0 = \pi/2$ yields the negative initial velocity $v(0) < 0$.

Similarly, for the initial condition that the point mass is at a turning point, we can find the initial velocity as follows.

$$\phi_0 = 0, \quad \xi(0) = A\cos(0) = A, \quad v(0) = -\omega A \sin(0) = 0 \tag{2.8}$$

$$\phi_0 = \pi, \quad \xi(0) = A\cos(-\pi) = -A, \quad v(0) = -\omega A \sin(-\pi) = 0 \tag{2.9}$$

The situation is the same for any value of $\phi_0$. It presents the initial position and velocity. Thus, the initial phase $\phi_0$ precisely describes the initial oscillatory motion.

**Natural Frequency**

The last parameter $\omega$ is determined by the physical property of the system. Hence, we need to get the equation of motion involved to find its explicit form.

Substitution of solution (2.4) into the equation of motion (the governing differential equation) (2.3) yields the following equality.

$$- \omega^2 m A \cos(\omega t + \phi_0) = -k A \cos(\omega t + \phi_0) \tag{2.10}$$

Cancellation of common terms from both sides of (2.10) leads to the following equation.

$$\omega = \sqrt{\frac{k}{m}} \equiv \omega_0 \tag{2.11}$$

Expression (2.11) indicates that the oscillation frequency $\omega$ is determined by the physical system, the spring constant, and the mass. Since the oscillation frequency is a constant for a given pair of $k$ and $m$, often we put a subscript 0 (to distinguish it from the driving frequency discussed later) and call this frequency $\omega_0$ the natural frequency of the spring-mass system.

We can intuitively comprehend the form of (2.11). The stronger the spring (the higher the $k$) or lighter the mass (the lower the $m$), the oscillation is faster, and hence $\omega$ is higher.

**Generality of** $A \cos(\omega t + \pi_0)$ **Solution**

As we speculated above, a sine function should satisfy the same equation of motion. Keeping this in mind, use $\phi_0 = -\pi/2$ in (2.4) and the following mathematical identity.

$$\cos(\omega t - \pi/2) = \sin(\omega t) \tag{2.12}$$

The right-hand side of (2.12 ) indicates that the same physical situation (that the point mass is at the equilibrium at $t = 0$ swinging toward the positive turning point) corresponds to $\phi_0 = 0$ if we employ a sine function. Of course, the condition that the point mass is at the positive turning point at $t = 0$ yields a different initial phase if we employ the sine function.

$$\xi(t) = A \sin(\omega t + \pi/2), \quad \xi(0) = A \sin(\pi/2) = A \tag{2.13}$$

Remembering that the sine and cosine functions are different only by the phase shift of $\pi/2$, we can comprehend the generality of the cosine solution (2.4) intuitively.

### 2.2.2 Visual Solution

Above, we solved the equation of motion governing a spring-point-mass system intuitively. Knowing that a sine or cosine function represents a harmonic oscillation, we used it as a test solution. In this section, we view the equation of motion as a linear differential equation and solve it numerically. In doing so, we visualize the solution by using a spreadsheet. This method clarifies the numerical meaning of initial conditions.

First, view the equation of motion (2.3) from the numerical point of view. To facilitate the numerical steps we divide the entire equation by mass $m$.

$$\frac{d^2 \xi(t)}{dt^2} = -\frac{k}{m} \xi(t) \equiv a\xi(t) \tag{2.14}$$

Here $k/m$ is the square of the natural frequency $\omega_0$ defined by (2.11). Since it is a material constant, we put it constant $a = -k/m$ on the right-hand side.

Equation (2.14) tells us that the second-order derivative of function $\xi(t)$ is proportional to the function itself with the constant of proportionality $a$. Defining $\xi'$ as the first order derivative of $\xi(t)$ as $\xi' = d\xi/dt$, we find the following expression for the differential $d\xi$.

$$d\xi(t) = \xi'(t)dt \tag{2.15}$$

Since the second-order derivative $d^2\xi/dt^2$ is the first-order derivative of $\xi'$, from (2.14) we find

$$\frac{d\xi'}{dt} = a\xi \tag{2.16}$$

hence,

$$d\xi'(t) = a\xi(t)dt \tag{2.17}$$

Graphically, we can view $d\xi'$ as the rise of the curve representing function $\xi(t)$ corresponding to the run $dt$ from $t$ to $t + dt$. Figure 2.2 illustrates this situation schematically.

Referring to Fig. 2.2, we note that the value of function $\xi'$ at $t = t + dt$ can be expressed as follows.

$$\xi'(t + dt) = \xi'(t) + d\xi'(t) \tag{2.18}$$

Using (2.17) we can express the right-hand side of (2.18) as follows.

$$\xi'(t + dt) = \xi'(t) + a\xi(t)dt \tag{2.19}$$

Expression (2.19) tells us that if we know $\xi(t)$ and $\xi'(t)$, we can find $\xi'(t + dt)$.

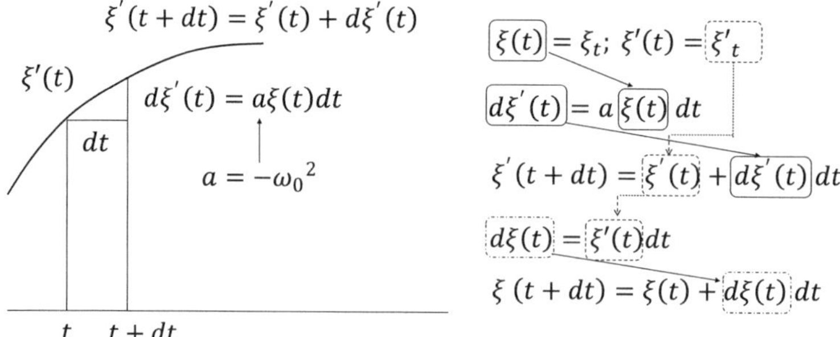


**Fig. 2.2** Conceptual illustration of visual solution. The equations on the right of the graph indicate that we can find $\xi(t + dt)$ and $\xi'(t + dt)$ from $\xi(t)$ and $\xi'(t)$

Once we know $\xi'(t + dt)$ by (2.19), we can find $\xi(t + dt)$ by using the following general expression.

$$\xi(t + dt) = \xi(t) + d\xi(t) = \xi(t) + \xi'(t)dt \tag{2.20}$$

Expressions (2.19) and (2.20) let us determine $\xi(t + dt)$ based on the knowledge of $\xi(t)$ and $\xi'(t)$. In other words, knowing $\xi$ and its derivative $\xi'$ at a time step $t$, we can find $\xi'$ and $\xi$ at the next time step $t + dt$. The right half of Fig. 2.2 illustrates these processes.

The above discussion indicates that if we know the initial displacement $\xi(0) \equiv \xi_0$ and initial velocity $v(0) = \xi'(0) \equiv v_0$, we can find the displacement at any time using the following recurrence relations.

$$
\begin{aligned}
\xi'_0 &= v_0; \ \ \xi_0 = \xi(0) \\
\xi'_1 &= \xi'_0 + a\xi_0 \Delta t; \ \ \xi_1 = \xi_0 + \xi'_0 \Delta t \\
&\quad \cdots \quad ; \quad \cdots \\
\xi'_{i+1} &= \xi'_i + a\xi_i \Delta t; \ \ \xi_{i+1} = \xi_i + \xi'_i \Delta t \\
&\quad \cdots \quad ; \quad \cdots \\
\xi'_n &= \xi'_{n-1} + a\xi_{n-1}\Delta t; \ \ \xi_n = \xi_{n-1} + \xi'_{n-1}\Delta t
\end{aligned}
\tag{2.21}
$$

Here $\xi_n$ and $\xi'_n = v_n$ represent the displacement and velocity at arbitrary time $t = n\Delta t$. Figure 2.3 shows a sample spreadsheet that implements the above recursive computation. In this example, the function $f(t)$ corresponds to the displacement $\xi(t)$, and we use the initial conditions $f(0) = 0$, $f'(0) = 1$, and the constant $a = -2$. To the right of the spreadsheet, the plot shows $f(t)$ for the first several seconds.

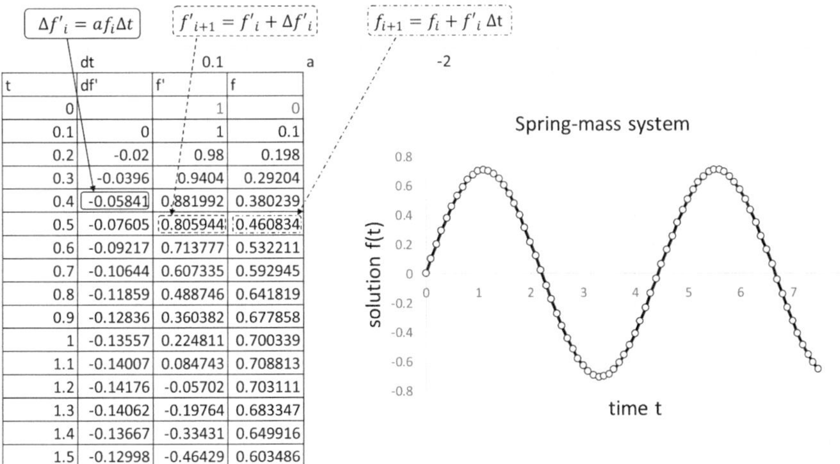


| t | df' | f' | f |
|---|---|---|---|
| 0 | | 1 | 0 |
| 0.1 | 0 | 1 | 0.1 |
| 0.2 | -0.02 | 0.98 | 0.198 |
| 0.3 | -0.0396 | 0.9404 | 0.29204 |
| 0.4 | -0.05841 | 0.881992 | 0.380239 |
| 0.5 | -0.07605 | 0.805944 | 0.460834 |
| 0.6 | -0.09217 | 0.713777 | 0.532211 |
| 0.7 | -0.10644 | 0.607335 | 0.592945 |
| 0.8 | -0.11859 | 0.488746 | 0.641819 |
| 0.9 | -0.12836 | 0.360382 | 0.677858 |
| 1 | -0.13557 | 0.224811 | 0.700339 |
| 1.1 | -0.14007 | 0.084743 | 0.708813 |
| 1.2 | -0.14176 | -0.05702 | 0.703111 |
| 1.3 | -0.14062 | -0.19764 | 0.683347 |
| 1.4 | -0.13667 | -0.33431 | 0.649916 |
| 1.5 | -0.12998 | -0.46429 | 0.603486 |

**Fig. 2.3** Sample spreadsheet to compute recursive relations to solve the equation of motion for a spring-mass system

## 2.3 Spring-Mass-Damper System

When the oscillation system contains a velocity-damping mechanism, we need to consider damping force. Consequently, the total external force expression and the equation of motion take the following forms.

$$f_{tot} = -k\xi(t) - b\frac{d\xi}{dt} \tag{2.22}$$

$$m\frac{d^2\xi(t)}{dt^2} + b\frac{d\xi}{dt} + k\xi(t) = 0 \tag{2.23}$$

Here $b$ is the damping coefficient. In (2.23), we move the force terms to the left of the equal sign to facilitate the discussion below.

As is the case of the spring-mass system where the spring force is the only external force on the point mass, a solution to the equation of motion (2.23) is oscillatory. Thus we can expect that (2.23) yields a sine and cosine function-like solution. In this case, however, the property of the cosine or sine function that the second-order derivative is proportion to the original function with a negative constant does not work. This is because the second term on the left-hand side of (2.23) is the first-order derivative of $\xi(t)$. If we start with $\cos(\omega t + \phi_0)$ as a test function, this term is proportional to $-\omega \sin(\omega t + \phi_0)$. If we start with $\sin(\omega t + \phi_0)$, this term is proportional to $\omega \cos(\omega t + \phi_0)$. In any case, it is impossible to cancel the second-order derivative term (the first term), the first-order derivative term (the second term), and the original function term (the third) term on the left-hand side of (2.23) with one another. We can speculate that a mixture of a cosine and sine function may satisfy the differential equation (2.23).

Well, we know a convenient tool to combine cosine and sine functions of the same argument (the phase term); Euler's notation. So, let's use the following function as a test solution to (2.23), and see how it goes.

$$\xi(t) = Ae^{i(\omega t + \phi_0)} \tag{2.24}$$

Substitute (2.24) into the equation of motion (2.23) and obtain the following equation.

$$\left(-m\omega^2 + ib\omega + k\right) Ae^{i(\omega t + \phi_0)} = 0$$

Since $Aexp(i(\omega t + \phi_0)) \neq 0$, we can divide the above equation by this term and obtain the following equation after arranging the terms to separate to the real and imaginary parts.

$$k - m\omega^2 + ib\omega = 0 \tag{2.25}$$

For (2.25) to hold, both the real and imaginary parts must be equal to 0. It follows that $\omega$ is a complex number. So, we set it as follows.

$$\omega = \omega_r + i\omega_i \tag{2.26}$$

Here $\omega_r$ and $\omega_i$ are real.

Substituting (2.26) into (2.25), we obtain the following set of equations.

$$k - m\omega_r{}^2 + m\omega_i{}^2 - b\omega_i = 0 \tag{2.27}$$

$$-2im\omega_r\omega_i + ib\omega_r = 0 \tag{2.28}$$

Solving (2.28) for $\omega_i$ and substituting the result into (2.27), we find the following expressions for the real and imaginary parts of $\omega$.

$$\omega_i = \frac{b}{2m} \tag{2.29}$$

$$\omega_r = \pm\sqrt{\frac{k}{m} - \frac{b^2}{4m^2}} \tag{2.30}$$

For simplicity, we introduce $\beta$ defined as follows and replace the first term inside the square root in (2.30) with the natural frequency $\omega_0$ defined by (2.11).

$$\beta = \frac{b}{2m} \tag{2.31}$$

Here $\beta$ is referred to as the decay constant as it indicates the decay characteristic of the oscillation. Below, we will see that $\beta$ explicitly indicates an exponential decay of an oscillation. Then we can rewrite (2.29) and (2.30) as follows.

$$\omega_i = \beta \tag{2.32}$$

$$\omega_r = \pm\sqrt{\omega_0^2 - \beta^2} \tag{2.33}$$

Using (2.32) and (2.33), we can express the solution (2.24) in the following form.

$$\xi(t) = Ae^{i\left(i\beta \pm \sqrt{\omega_0^2 - \beta^2}\right)t + i\phi_0}$$

$$= e^{i\phi_0}\left\{ A_1 e^{\left(-\beta + i\sqrt{\omega_0^2 - \beta^2}\right)t} + A_2 e^{\left(-\beta - i\sqrt{\omega_0^2 - \beta^2}\right)t} \right\} \tag{2.34}$$

Depending on the sign of the radicand $\omega_0^2 - \beta^2$, the solution $\xi(t)$ behaves differently. In the following sections, we discuss the three cases, the radicand is (1) negative, (2) zero, and (3) positive. These cases are called the over-damping, critical-damping, and under-damping, respectively.

**(1) Over damping ($\omega_0 < \beta$)**

In this case, the radicant is negative and therefore we can express $\sqrt{\omega_0^2 - \beta^2} = i\sqrt{\beta^2 - \omega_0^2}$, and rewrite (2.34) as follows.

$$\xi(t) = e^{i\phi_0} \left\{ A_1 e^{\left(-\beta - \sqrt{\beta^2 - \omega_0^2}\right)t} + A_2 e^{\left(-\beta + \sqrt{\beta^2 - \omega_0^2}\right)t} \right\} \tag{2.35}$$

$$v(t) = \dot{\xi}(t) = e^{i\phi_0} \left\{ A_1 \left(-\beta - \sqrt{\beta^2 - \omega_0^2}\right) e^{\left(-\beta - \sqrt{\beta^2 - \omega_0^2}\right)t} \right\}$$

$$+ e^{i\phi_0} \left\{ A_2 \left(-\beta + \sqrt{\beta^2 - \omega_0^2}\right) e^{\left(-\beta + \sqrt{\beta^2 - \omega_0^2}\right)t} \right\} \tag{2.36}$$

Since $0 < \beta^2 - \omega_0^2 < \beta^2$, the exponents of the two exponential functions are negative; $-\beta - \sqrt{\beta^2 - \omega_0^2} < 0$ and $-\beta + \sqrt{\beta^2 - \omega_0^2} < 0$. Therefore, (2.35) represents an exponentially decaying function of $t$. In other words, the displacement decays with time without showing an oscillatory behavior. The explicit form of the solution can be found if we impose initial conditions. Here, we consider the following initial conditions; the point mass is at 1 (in an arbitrary unit) away from the equilibrium and at rest at $t = 0$.

$$\xi(0) = 1 \tag{2.37}$$

$$v(0) = \dot{\xi}(0) = 0 \tag{2.38}$$

Substitution of (2.37) and (2.38) into (2.35) and (2.36) yields the following set of equations.

$$e^{i\phi} (A_1 + A_2) = 1 \tag{2.39}$$

$$e^{i\phi} \left\{ A_1 \left(-\beta - \sqrt{\beta^2 - \omega_0^2}\right) + A_2 \left(-\beta + \sqrt{\beta^2 - \omega_0^2}\right) \right\} = 0 \tag{2.40}$$

Setting the initial phase $\phi_0 = 0$ (this setting does not cause loss of generality for the discussion here), we can find the amplitude $A_1$ and $A_2$ as follows.

$$A_1 = \frac{\beta + \sqrt{\beta^2 - \omega_0^2}}{2\sqrt{\beta^2 - \omega_0^2}} \tag{2.41}$$

$$A_2 = \frac{-\beta + \sqrt{\beta^2 - \omega_0^2}}{2\sqrt{\beta^2 - \omega_0^2}} \tag{2.42}$$

Thus, we can determine an explicit form of the solution under the overdamping condition. Figure 2.4a illustrates a sample of overdamping characteristics.

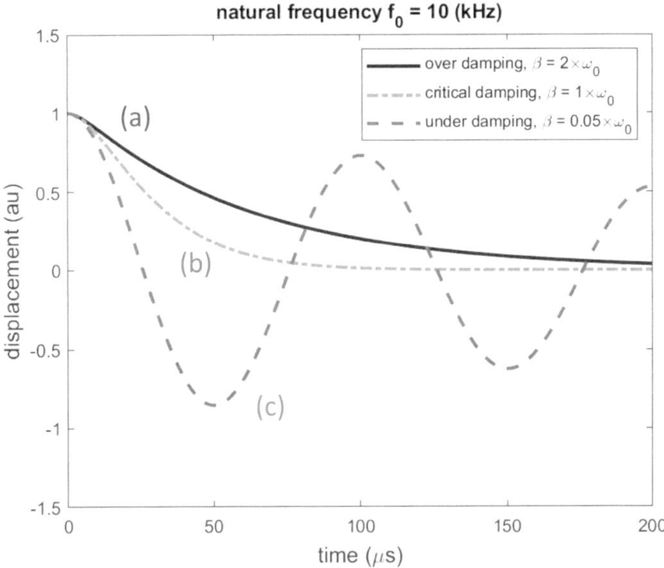



**Fig. 2.4** **a** Over-damping; **b** critical-damping; and **c** under-damping characteristics of point-mass oscilation

(2) Critical damping ($\omega_0 = \beta$)

In this case, the radicant is zero, hence from (2.34) $\xi(t)$ takes the following form.

$$\xi(t) = Be^{-\beta t}$$

Since the differential equation (2.23) is second order, we need one more term in the general solution. It follows that $B$ is a function of time. (Because if it is a constant, we cannot make another term.) It is easily proved that the following form of solution can satisfy the differential equation [4].

$$\xi(t) = B_1 e^{-\beta t} + B_2 t e^{-\beta t} \tag{2.43}$$

Here is the proof.

$$\dot{\xi}(t) = -B_1 \beta e^{-\beta t} + B_2 e^{-\beta t} - B_2 \beta t e^{-\beta t} \tag{2.44}$$

$$\ddot{\xi}(t) = B_1 \beta^2 e^{-\beta t} - 2B_2 \beta e^{-\beta t} + B_2 \beta^2 t e^{-\beta t} \tag{2.45}$$

Substitute (2.44) and (2.45) into (2.23) using (2.31) and divide the resultant equation by the non-zero term $exp(-\beta t)$. Then we obtain the following equation.

$$
\begin{aligned}
m\ddot{\xi} + b\dot{\xi} + k\xi \\
= m \left\{ \left( B_1 \beta^2 - 2B_2 \beta + B_2 \beta^2 t \right) + 2\beta \left( -B_1 \beta + B_2 - B_2 \beta t \right) \right\} + k \left( B_1 + B_2 t \right) \tag{2.46}
\end{aligned}
$$

From (2.11) and the critical damping condition $\omega_0 = \beta$, we find the following relationships.

$$k = m\omega_0^2 = m\beta^2 \tag{2.47}$$

Using (2.47) in the last term on the left-hand side, we can write (2.46) as follows.

$$m\left(B_1\beta^2 - 2B_2\beta + B_2\beta^2 t\right) + 2m\beta\left(-B_1\beta + B_2 - B_2\beta t\right) + m\beta^2\left(B_1 + B_2 t\right)$$
$$= m\left\{\left(B_1\beta^2 - 2B_2\beta - 2\beta^2 B_1 + 2\beta B_2 + \beta^2 B_1\right) + \left(B_2\beta^2 - 2\beta B_2\beta + \beta^2 B_2\right)\right\} t \tag{2.48}$$

Since the content of the two parentheses is zero, (2.48) is zero for all $t$. Thus, the left-hand side of (2.46) is zero, which means (2.43) satisfies the equation of motion under the critical damping condition.

q.e.d.

Determine the coefficients $B_1$ and $B_2$ using the same initial conditions as the over damping case. Substituting (2.43) into (2.37) and (2.38), we obtain the following set of equations.

$$B_1 \quad = 1 \tag{2.49}$$
$$-\beta B_1 + B_2 = 0 \tag{2.50}$$

Thus, we find the general solution for the critical damping case in the following form.

$$\xi_{cd} = e^{-\beta t} + \beta t e^{-\beta t} = (1 + \beta t)e^{-\beta t} \tag{2.51}$$

(3) Under damping ($\omega_0 > \beta$)

This case is known as the under-damping or decaying oscillation and is most important in oscillation dynamics. Putting $\omega = \sqrt{\omega_0^2 - \beta^2}$ we can express the exponent of the terms in the curly brackets on the right-hand side of (2.35), denoted by $\lambda$, in the following form.

$$\lambda = -\beta \pm \sqrt{\beta^2 - \omega_0^2} = -\beta \pm i\sqrt{\omega_0^2 - \beta^2} = -\beta \pm i\omega \tag{2.52}$$

With expression (2.52), the general solution for the under-damping case becomes as follows.

$$\xi_{ud} = e^{-\beta t}\left(C_1 e^{i\omega t} + C_2 e^{-i\omega t}\right)$$
$$= e^{-\beta t}\left\{C_1\left(\cos\omega t + i\sin\omega t\right) + C_2\left(\cos\omega t - i\sin\omega t\right)\right\}$$
$$= e^{-\beta t}\left\{(C_1 + C_2)\cos\omega t + (C_1 - C_2)i\sin\omega t\right\}$$
$$= C_0 e^{-\beta t}\cos(\omega t - \delta_0) \tag{2.53}$$

Here $C_0$ and $\delta_0$ are determined with initial conditions. Again, Substituting (2.53) into (2.37) and (2.38), we obtain the following set of equations.

$$C_0 \cos(\delta_0) = 1 \tag{2.54}$$
$$-\beta \cos\delta_0 + \omega \sin\delta_0 = 0 \tag{2.55}$$

With conditions (2.54) and (2.55), we can write the general solution in this case as follows.

$$\xi_{ud} = \frac{1}{\cos \delta_0} e^{-\beta t} \cos(\omega t - \delta_0) \tag{2.56}$$

where

$$\omega = \sqrt{\omega_0^2 - \beta^2} \tag{2.57}$$

$$\delta_0 = \tan^{-1}\left(\frac{\beta}{\omega}\right) \tag{2.58}$$

Above, we discussed that $\omega_0$ and $\beta$ represent the natural frequency and decay constant of an oscillation. Here (2.56) and (2.57) explicitly indicate the meaning of these two quantities. According to (2.56), an under-damped wave has the decaying feature expressed by the exponential term of the expression and the oscillatory feature expressed by the cosine term. The decay term indicates that the oscillation decays at a rate of $\beta$; the wave's amplitude decreases by $exp(-\beta)$ in unit time. The (angular) frequency of the oscillatory term (2.57) becomes equal to $\omega_0$ when the decay constant is null. In other words, the natural frequency is the oscillation frequency when the system does not have a mechanism to dissipate the oscillation energy.

Further, (2.58) indicates that when the system undergoes decay, not only the oscillation frequency deviates from the natural frequency with an increase in the decay constant but also the phase is delayed by $\delta_0$. According to (2.58), when the decay constant is zero, $\delta_0 = 0$ and there is no delay from the decay-free oscillation.

Figure 2.5 shows the decaying characteristics of two sample under-damped oscillations. The natural frequency of both samples is commonly $f_0$=10 kHz ($\omega_0 = 2\pi f_0 = 20\pi$ krad/s). The decay constant of the more slowly decaying oscillation is one-hundredth of the natural angular frequency ($\beta = \omega_0/100$), and that of the faster decaying one is one-twentieth of the natural angular frequency ($\beta = \omega_0/20$). This type of oscillatory decaying characteristic is called the ring-down characteristic.

According to (2.57) the frequency shift for the faster decaying case is $\omega_0 \to \omega = \sqrt{\omega_0^2 - (\omega_0/20)^2} = 0.9975\omega_0$, and for the other case is $\omega_0 \to \omega = \sqrt{\omega_0^2 - (\omega_0/100)^2} = 0.9999\omega_0$. In either case, the frequency shift is of the order of a tenth of a percent or less. Similarly, the phase shift due to the decay characteristic is small. According to (2.58), the phase shifts for the faster and slower decay cases are, respectively, $\tan^{-1}(\beta/\omega) = 0.0501$ (rad) i.e., 0.8% of one period, and $\tan^{-1}(\beta/\omega) = 0.01$ (rad) i.e., 0.16% of one period.

On the other hand, the decay characteristics are significant. The time constant (the time to decay by a factor of $e$, the base of natural logarithms) is $1/\beta = 20/\omega_0 = (20/(2\pi f_0)) = 3.2/f_0 = 3.2\tau$ for the faster decaying case, and $1/\beta = 100/\omega_0 = (100/(2\pi f_0)) = 15.9/f_0 = 15.9\tau$ for the slowly decaying case. Here $\tau = 1/f_0$ is the oscillation period. These numbers indicate that the signal decreases by $1/e$ in 3.2 oscillations and 15.9 oscillations, respectively. Figure 2.5 illustrates that for every $15.9 \approx 16$ oscillations the peak value of the slowly decaying oscillation decreases by $e$. It is important to note that this characteristic is

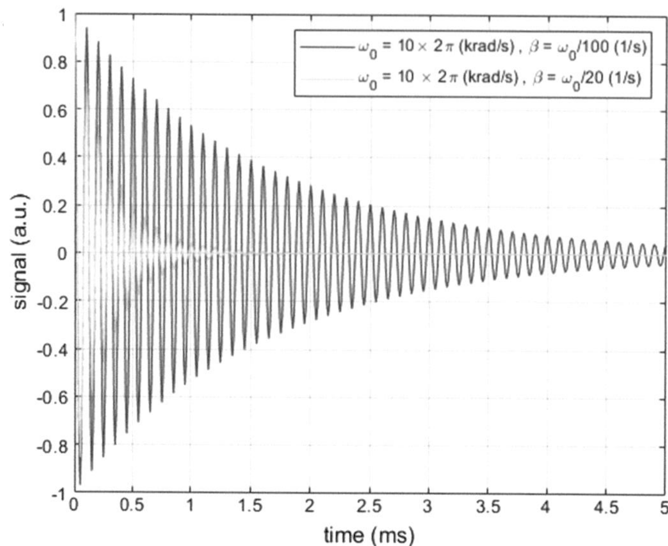



**Fig. 2.5** Ring down characteristics of two samples. Natural frequency is 10 kHz commonly to the two samples whereas the decay constants are $\omega_0/100$ (the slower decay case) and $\omega_0/20$ (the faster decay case.)

the same whichever peak you start to count 16 oscillations, starting at the peak near $t = 0$ (s) to the $16^{\text{th}}$ peak near $t = 1.6$ (ms) or starting at $t = 1.5$ (s) to $t = 3.1$ (s). This feature is convenient to determine the time constant of a physical system, e.g., a spring and damper system, experimentally. All you need to do is to measure the amplitude at a certain time (called $t_1$) and find the other time ($t_2$) when the amplitude decreases by a factor of $e$. $t_2 - t_1$ is the time constant of the decaying signal and its reciprocal $1/(t_2 - t_1) = \beta$ is the decay constant of the physical system, e.g., the damper.

We can solve the equation of motion for a spring-mass-damper system using the spreadsheet in the same fashion as we discussed above for the spring-mass system. In the present case, we need to add the damping force term in the expression of $d\xi'$. From (2.23), we find the following expression of $d\xi'$, which corresponds to (2.17) in the spring-mass (with no damping) case.

$$d\xi'(t) = (a\xi(t) - 2\beta v(t))\, dt \tag{2.59}$$

Here $a = -\omega_0^2$, $\beta$ is as defined by (2.31), and $v$ is the velocity or the first-order derivative of $\xi(t)$. Figure 2.6 illustrates this procedure in the same fashion as Fig. 2.3. In Fig. 2.6, we use $a = -2$ and $2\beta = 0.2$.

**Complex Spring Constant**

The above under-damping dynamics is conveniently represented with a complex spring constant. Rewrite the equation of motion (2.3) by replacing the real spring constant $k_{sp}$ with a complex spring constant $k_{sp} \rightarrow k_r + i k_i$.

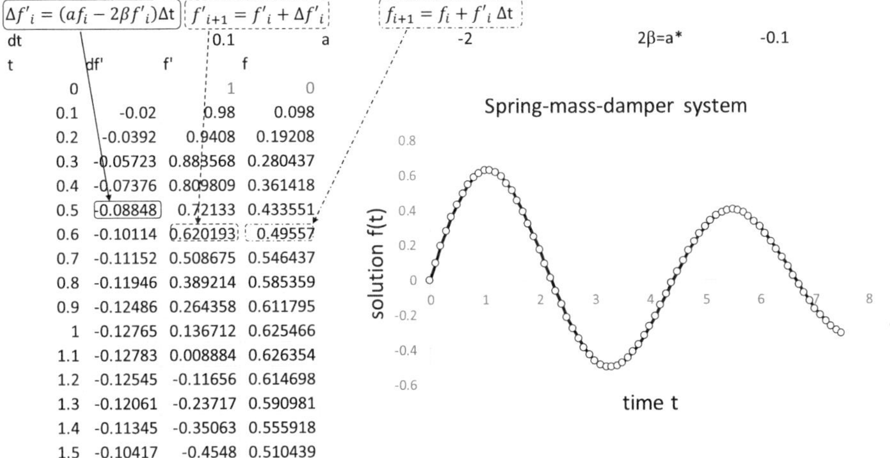


**Fig. 2.6** Sample spreadsheet to compute recursive relations to solve the equation of motion for a spring-mass-damper system

$$m\frac{d^2\xi(t)}{dt^2} = -(k_r + ik_i)\xi(t) \tag{2.60}$$

Here we prove that the solution in the following form satisfies the equation of motion (2.60).

$$\xi(t) = Ae^{\left(-\beta \pm i\sqrt{\omega_0^2 - \beta^2}\right)t} \tag{2.61}$$

Note that (2.61) is the exponential form of the solution we derived for the under-damping case (See (2.56) and (2.57)). Substitution of (2.61) into (2.60) leads to the following equations.

$$\frac{k_r}{m} = \omega_0^2 - 2\beta^2 \tag{2.62}$$

$$\frac{k_i}{m} = \pm 2\beta\sqrt{\omega_0^2 - \beta^2} \tag{2.63}$$

Remembering that $\omega_0^2 = k/m$ (2.11) and $2\beta = b/m$ (2.31), we find that the real part of the complex spring constant $k_r$ is associated with $\omega_0^2 = k/m$ (2.11), and the imaginary part $k_i$ is associated with the damping coefficient $b$. (2.62) represents that the damping effect reduces the spring constant $k$ as much as $2\beta^2$. We can interpret (2.62) and (2.63) as follows. As the damping effect ($\beta$) is included in the real part $k_r$, the imaginary part $k_i$ is weakened by the factor $\sqrt{\omega_0^2 - \beta^2}$. In other words, in the complex spring constant formalism, the damping and oscillatory (elastic) effects are distributed in both the real and imaginary parts of the spring constant.

## 2.4 Forced Oscillation System

The dynamics of the spring-mass system and spring-mass-damper system are often called unforced oscillations because no external agent exerts a force on the mass. When an external agent drives the object connected to a spring, the resultant oscillation is called forced oscillation. It is especially important to understand oscillation when the external agent applies harmonic force on the object. This type of forced oscillation is called Damped Driven Harmonic Oscillation and the behavior of the forced oscillated object is referred to as the object's harmonic response. It may sound like Damped Driven Harmonic Oscillation is a special case of forced oscillation. However, this is not necessarily the case. Remember Fourier's theorem. Most realistic functions can be expanded to a Fourier series. Therefore, analysis of harmonic response can be viewed as a frequency component of the object's response to a given type of force.

### 2.4.1 Damped Driven Harmonic Oscillation

We can derive the equation of motion for a Damped Driven Harmonic Oscillation generally by replacing the zero on the right-hand side of the equation of motion for unforced case (2.23) with a harmonic driving force.

$$m\frac{d^2\xi(t)}{dt^2} = -k_{sp}\xi(t) - b\frac{d\xi(t)}{dt} + f_{ex}\sin\Omega t \tag{2.64}$$

This leads to the following non-homogeneous differential equation.

$$\frac{d^2\xi(t)}{dt^2} + 2\beta\frac{d\xi(t)}{dt} + \omega_0^2\xi(t) = \frac{f_{ex}}{m}\sin\Omega t \tag{2.65}$$

The homogeneous solution to (2.65) is the same as the general solution to the case of unforced oscillation with a damping mechanism. In engineering, normally particular solutions are more important. So below, we discuss the particular solution.

Consider the following test solution. In the case of unforced oscillation, the oscillatory solution (the under-damping case) has an exponential decay term multiplied to the sinusoidal term (2.56). You may wonder why we do not use the exponential term this time. The reason will be clarified shortly in this section.

$$\xi(t) = A\sin(\Omega t - \delta) \tag{2.66}$$

Substitution of (2.66) into (2.65) yields the following equation.

$$-\Omega^2\sin(\Omega t - \delta) + 2\beta\Omega\cos(\Omega t - \delta) + \omega_0^2\sin(\Omega t - \delta) = \frac{f_{ex}}{Am}\sin\Omega t \tag{2.67}$$

By expanding $\sin(\Omega t - \delta)$ and $\cos(\Omega t - \delta)$ terms, we can rewrite (2.67) into the following form.

$$\left( (\omega_0^2 - \Omega^2) \cos\delta + 2\beta\Omega \sin\delta - \frac{f_{ex}}{Am} \right) \sin(\Omega t)$$
$$+ \left( -(\omega_0^2 - \Omega^2) \sin\delta + 2\beta\Omega \cos\delta \right) \cos\Omega t = 0 \tag{2.68}$$

For (2.68) to hold for all time $t$, the content of the parentheses in front of the $\sin\Omega t$ and $\cos\Omega t$ terms must be respectively zero.

$$(\omega_0^2 - \Omega^2) \cos\delta + 2\beta\Omega \sin\delta = \tfrac{f_{ex}}{Am} \tag{2.69}$$
$$2\beta\Omega \cos\delta - (\omega_0^2 - \Omega^2) \sin\delta = 0 \tag{2.70}$$

By solving (2.71) and (2.72) for $\cos\delta$ and $\sin\delta$, we obtain the following equations.

$$\cos\delta = \frac{f_{ex}}{Am} \frac{\omega_0^2 - \Omega^2}{(\omega_0^2 - \Omega^2)^2 + (2\beta\Omega)^2} \tag{2.71}$$
$$\sin\delta = \frac{f_{ex}}{Am} \frac{2\beta\Omega}{(\omega_0^2 - \Omega^2)^2 + (2\beta\Omega)^2} \tag{2.72}$$

It follows that the amplitude $A$ and the phase $\delta$ are expressed as follows.

$$A = \frac{f_{ex}/m}{\sqrt{\left(\omega_0^2 - \Omega^2\right)^2 + (2\beta\Omega)^2}} \tag{2.73}$$
$$\tan\delta = \frac{2\beta\Omega}{\omega_0^2 - \Omega^2} \tag{2.74}$$

As long as (2.73) and (2.74) hold, solution (2.66) satisfies the equation of motion (2.65). We can see that with an increase of $\beta$, i.e., the damping effect, the amplitude $A$ reduces, and the phase shift from the driving force $\delta$ increases.

### 2.4.2  Amplitude and Phase of Forced Oscillation

Equations (2.73) and (2.74) indicate that the amplitude and phase of Damped Driven Harmonic Oscillation depend on the driving frequency and decay constant. Figure 2.7a plots the displacement of the spring-mass system represented by mass 1 kg, natural frequency 500 Hz, and decay constant 40 1/s and 100 1/s. It is seen that the higher decay constant lowers the amplitude of the displacement. Figure 2.7b compares the same oscillatory system when driven at 500 Hz, 475 Hz and 525 Hz. It is seen that as the driving frequency deviates from the natural frequency, the amplitude decreases. It is also seen that the phase delay (the phase difference between the driving force and the mass's displacement $\delta$) varies depending on the driving frequency, as expressed by (2.74).

Figure 2.7 indicates that the oscillation frequency is the same as the driving frequency regardless of the natural frequency and that an increase in the damping coefficient makes the oscillation amplitude smaller. The former literally indicates that the oscillation is "forced" by the driving force. The latter reflects the fact that the oscillation is a particular solution and the increase in the damping coefficient causes the energy transfer from the driving mechanism to the oscillation less efficient (because part of the driving energy is dissipated). This is contrastive to the case of unforced oscillation where an increase in the damping coefficient simply makes the oscillator ring down more quickly. The exact same argument holds for the forced oscillation of a simple pendulum as (2.102) and (2.103) indicate.

Figure 2.8 plots (2.73) and (2.74) as a function of driving frequency for $f_{ex} = 1$ (N), $m = 1$ (kg), $\omega_0 = 1000\pi$ (rad/s), and $\beta = 40, 100$ (1/s). In the experimental sense, each point of the plot (a) indicates how largely the mass oscillates when the external force drives the system with a harmonic function at the corresponding frequency shown on the horizontal

**Fig. 2.7** Time trend of the displacement of the linear system represented by mass 1 kg, natural frequency 500 Hz, and decay constant 40 1/s when driven at 500 Hz, 475 Hz, and 525 Hz

**Fig. 2.8** Amplitude and phase vs driving frequency of linear system represented by mass 1 kg, natural frequency 500 Hz, and decay constant 40 1/s and 100 1/s

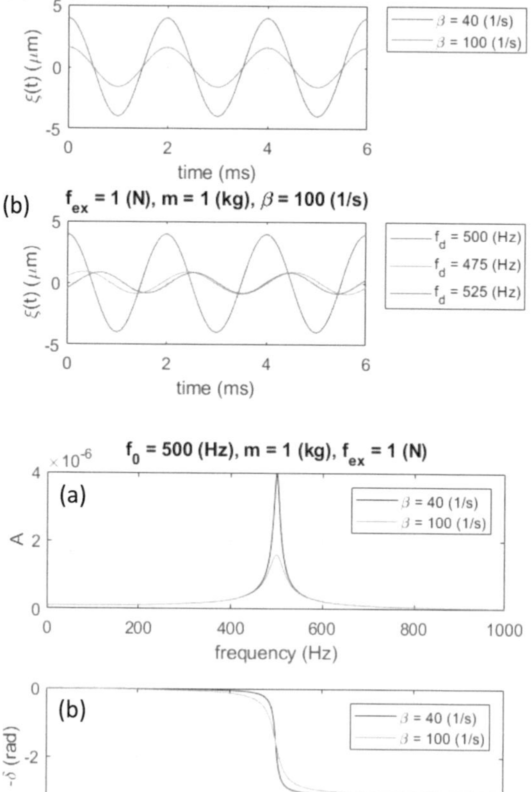

axis. The entire curve of Fig. 2.8a represents the variation of the magnitude of the oscillation as a function of the driving frequency. Notice that the higher the decay constant lower the peak and broader the spectrum. This phenomenon is known as the spectral broadening due to energy dissipation. (Sometimes we say that the dissipation mechanism lowers the quality (Q) factor of the oscillation.) We can say that it is the frequency version of impulse response discussed in Sect. 6.3. Similarly, Fig. 2.8b represents the frequency dependence of the phase.

Notice that when the driving frequency is equal to or close to the natural frequency, the oscillatory system shows strong dependence on the driving frequency. The condition that the driving frequency is equal to the natural frequency is referred to as resonance. At or near resonance frequency, the oscillatory system exchanges energy with the driving mechanism most efficiently. Often analysis of the system's behavior at or near resonance frequency is important to characterize the oscillatory system. We will revisit these issues in Sect. 6.6.2.

## 2.5    Examples of Harmonic Oscillations

### 2.5.1    Vertical Spring Oscillation

As another example, we derive the equation of motion for a mass connected to a vertical spring and solve it. Unlike the horizontal spring system discussed in Fig. 2.1, gravity exerts a static (constant) downward force on the mass. Does this static force affect the oscillation? It is not a straightforward question to answer. We will consider it in the following discussion. (For simplicity the damping term is not included in the discussion here. The damping effect does not change the argument on the effect of the static force.)

Consider in Fig. 2.9 a point mass $m$ is hanging from the lower end of a spring whose spring constant is $k$. The equation of motion is as follows.

$$m \frac{d^2 y(t)}{dt^2} = -k \left( y(t) + y_0 \right) + mg \tag{2.75}$$

Here the positive direction of the $y$-axis is downward, $y$ is the displacement from the point where the spring is at its natural length, and $y_0$ is the vertical location where upward spring force balances the downward gravity on the point mass. Naturally,

$$y_0 = \frac{mg}{k} \tag{2.76}$$

Substitution of (2.76) into (2.75) yields the following differential equation.

$$m \frac{d^2 y(t)}{dt^2} = -k \left( y(t) + y_0 \right) + mg = -ky(t) - ky_0 + mg = -ky(t) \tag{2.77}$$

Put the solution in the following form.

$$y(t) = A \sin(\omega t + \phi_0) \tag{2.78}$$

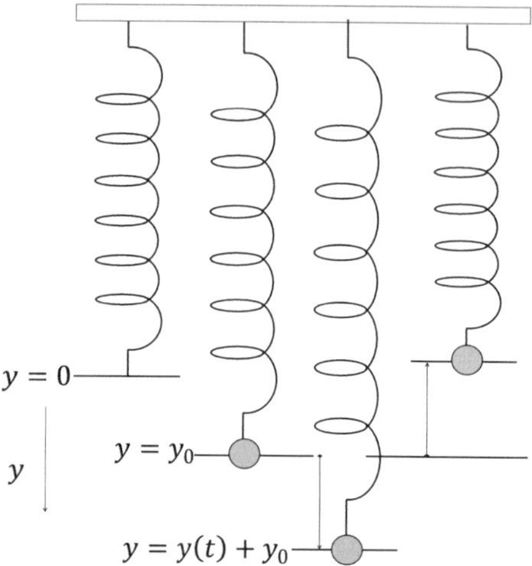

**Fig. 2.9** Vertical spring-mass system



Substitution of solution (2.78) into differential equation (2.77) indicates that the point mass oscillates with the following angular frequency.

$$\omega^2 = \frac{k}{m} \tag{2.79}$$

Since when the amplitude $A$ is null $y(t) = 0$, solution (2.78) indicates that the point mass oscillates around the equilibrium point with amplitude $A$. Use the following conditions.

$$y(0) = 1 \tag{2.80}$$

$$\dot{y}(0) = 0 \tag{2.81}$$

Substitution of (2.78) into (2.80) and (2.81) leads to the following solution.

$$A \sin(\phi_0) = 1 \tag{2.82}$$

$$A\omega \cos \phi_0 = 0 \tag{2.83}$$

Thus, we find $A = 1$ and $\phi_0 = \pi/2$, and the solution is as follows.

$$y(t) = \sin\left(\sqrt{\frac{k}{m}}t + \frac{\pi}{2}\right) \tag{2.84}$$

Note that the static force $mg$ shifts the equilibrium point from the end of the spring (when it is in the natural (unstretched) length) by $mg/k$ but it does not affect the oscillation frequency. The amplitude and the initial phase of the oscillation are determined by the two initial conditions (2.82) and (2.83), as is the case of a point mass connected to a horizontal spring. The oscillation frequency is the same as the natural frequency, which is determined by the spring constant and the mass. In other words, the system of spring and mass selects the oscillation frequency, regardless of the environment (the influence of gravity).

### 2.5.2  Pendulum

As another classic example of oscillation, consider a simple pendulum. Refer to Fig. 2.10 and consider the equation of motion for the horizontal and vertical dynamics.

$$m\frac{d^2x(t)}{dt^2} = -T\sin(\theta(t)) \tag{2.85}$$

$$m\frac{d^2y(t)}{dt^2} = T\cos(\theta(t)) - mg = 0 \tag{2.86}$$

Here $x$ and $y$ are the horizontal and vertical displacement of the bob from the equilibrium, $T$ is the tension exerted by the string, $\theta << 1$ is the angle between the string and a vertical line from the suspension point, and $m$ is the mass of the bob. In (2.86), we set the vertical acceleration to zero because angle $\theta$ is small.

Eliminating tension $T$ from (2.85) and (2.86), we obtain the following equation.

$$m\frac{d^2x(t)}{dt^2} = -mg\tan(\theta(t)) \tag{2.87}$$

Using the small angle approximation, we can put as follows.

$$\tan\theta(t) \simeq \theta(t) \tag{2.88}$$

$$x(t) \simeq l\theta(t) \tag{2.89}$$

**Fig. 2.10** Simple pendulum

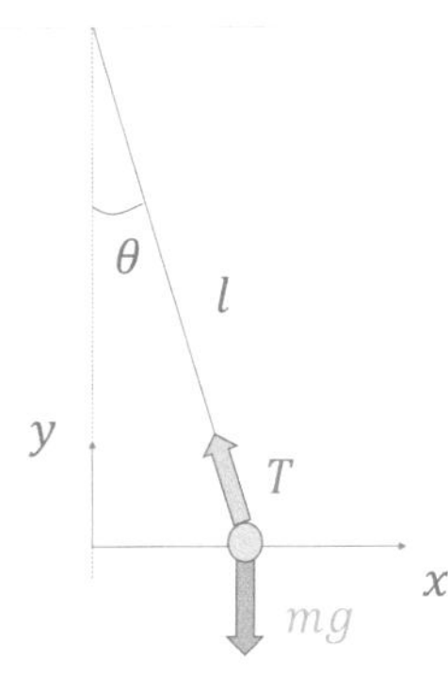

Substitution of (2.88) and (2.89) into (2.87) yields the following differential equation.

$$\frac{d^2 x(t)}{dt^2} = -\frac{g}{l} x(t) \tag{2.90}$$

Equation (2.90) is identical to the differential equation (2.23) for the spring-mass system if the ratio of the spring constant over mass $k/m = \omega_0^2$ is replaced with $g/l$ and the damping coefficient is set to zero, $b = 0$. Thus, by putting $b = 0$ in (2.53), we can immediately find the general solution to (2.90) in the following form.

$$x(t) = D_0 \cos\left(\sqrt{\frac{g}{l}} t - \delta_0\right) \tag{2.91}$$

Use the following initial conditions, as an example.

$$x(0) = 1 \tag{2.92}$$

$$\dot{x}(t) = 0 \tag{2.93}$$

We can determine the amplitude $D_0$ and phase $\delta_0$ by substituting (2.91) into initial conditions (2.92) and (2.93). Thus, we find as follows.

$$x(t) = \cos\left(\sqrt{\frac{g}{l}} t\right) \tag{2.94}$$

Notice that in this case, the gravity behaves completely differently from the case of the vertical spring-mass system discussed above. Here, it has a fundamental contribution to the oscillation. Its horizontal component is the driving force of the oscillation and that is why the gravitational acceleration is part of the frequency expression. For the same length, the oscillation frequency is lower on the moon than on the Earth because the gravity there is approximately one-sixth of the Earth.

### 2.5.3  Forced Oscillation of a Simple Pendulum

As another example of forced oscillation, consider exciting a simple pendulum at its suspension point in Fig. 2.11.

The equation of motion is the same as (2.87) and

$$m \frac{d^2 x(t)}{dt^2} = -mg \tan\theta \simeq -mg\theta \tag{2.95}$$

Here, unlike the unforced case, the angle $\theta(t)$ is related to the differential displacement of the bob relative to the suspension point.

$$x(t) - x_s(t) = l\theta(t) \tag{2.96}$$

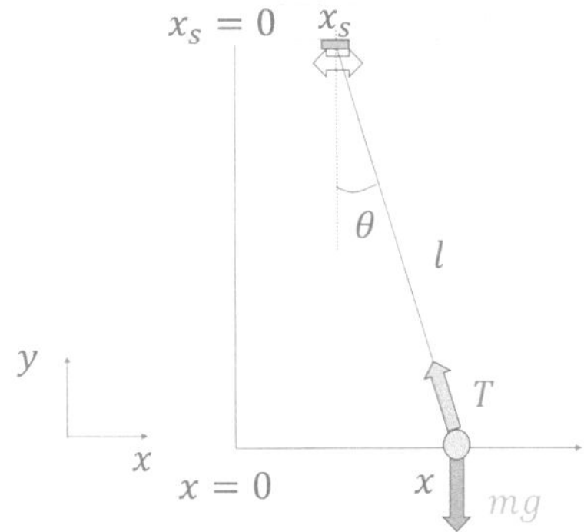

**Fig. 2.11** Simple pendulum oscillated at its suspension point

From (2.95) and (2.96), we obtain the following differential equation.

$$\frac{d^2x(t)}{dt^2} + \frac{g}{l}x(t) = \frac{g}{l}x_s(t) \tag{2.97}$$

Consider that the suspension point oscillates sinusoidally and the pendulum is subject to a certain damping mechanism represented by decay constant $\beta$.

$$\frac{d^2x(t)}{dt^2} + 2\beta\frac{dx(t)}{dt} + \frac{g}{l}x(t) = \frac{g}{l}X_s \sin \Omega t \tag{2.98}$$

Here, $X_s$ is the amplitude of the suspension point oscillation. The damping mechanism can be friction at the suspension point [5].

Equation (2.98) is identical to (2.65) with the following substitutions.

$$\omega_0^2 = \frac{g}{l} \tag{2.99}$$

$$\frac{f_{ex}}{m} = \omega_0^2 X_s \tag{2.100}$$

Expression (2.99) represents the natural angular frequency of the pendulum. The right-hand side of (2.100) can be interpreted as the amplitude of the acceleration of the suspension point. Thus, we find the general solution to (2.98) to take the same form as solution (2.66) for the spring-mass case.

$$\xi(t) = A \sin(\Omega t - \delta) \tag{2.101}$$

Repeating the same argument as the spring-mass case, we obtain the following expressions for the amplitude and phase.

$$A = \frac{\omega_0^2 X_s}{\sqrt{\left(\omega_0^2 - \Omega^2\right)^2 + (2\beta\Omega)^2}} \tag{2.102}$$

$$\tan \delta = \frac{2\beta\Omega}{\omega_0^2 - \Omega^2} \tag{2.103}$$

Notice that the denominator of (2.102) increases with the decay constant $\beta$. This indicates that the damping effect lowers the amplitude of the oscillation. This answers the above-raised question of "why an exponential term does not appear in solution (2.66)". In forced oscillation, the damping effect does not cause the steady-state oscillation to decay, but instead, reduces the efficiency of the driving oscillatory force transferred to the motion of the mass. (The general solution consists of a homogeneous solution and a particular solution. The former represents exponential decay and the latter the steady-state response of the system to the driving force. The homogeneous solution has an exponential decay term as is the case of the unforced oscillation case discussed above. However, in most practical cases, the steady-state oscillation is of main interest. So, here we do not include the transitional part in the solution.)

Above I used the word "transfer" in the relationship between the driving oscillatory force and the resultant oscillation of the mass. The meaning of this will become clear in a later section where we discuss the dynamics in the frequency domain. In the same section, we will discuss the effect of the decay constant on the phase delay seen in (2.103).

## References


1. James JF (2007) A student's guide to Fourier transforms: with applications in physics and engineering, 3rd edn. Cambridge University Press, Cambridge
2. Wiener N (1993) The Fourier integral and certain of its applications. Cambridge University Press, New York, USA (Reissued in digital print 2008)
3. Boas ML (2006) Mathematical methods in the physical sciences, 3rd edn. Wiley, London, Chap 7
4. Yoshida S (2017) Waves; Fundamental and dynamics. San Rafael, CA, USA, IOP Publishing, Bristol, UK, Morgan & Claypool
5. Yoshida S, Findley T (2005) Analysis of a simple pendulum driven at its suspension point. Euro J Phys 26(3):493–499

# Wave Equations and Solutions 3

When different parts of the material oscillate with phase delays, the oscillation is transferred as a wave. Single-point mass does not make a wave. Multiple point masses connected with springs can make a wave. Elastic media can be viewed as comprising atoms connected with springs. Thus, we can interpret the elastic waves propagating through an elastic medium as the motion of point-mass oscillation.

In this chapter, we first derive wave equations governing elastic waves as an extension of the harmonic oscillation we discussed in the preceding chapter and solve them. These solutions represent the so-called plane wave that illustrates several significant features of the wave dynamics. In the second part of this chapter, we discuss a more realistic wave known as the Gaussian wave. It is a wave localized around the axis of propagation. The basic properties of Gaussian waves are presented. Finally, we discuss some physical aspects of wave dynamics, i.e., Heygens' principle and dispersion.

## 3.1 Wave Equation

### 3.1.1 Transverse Wave on a String

Consider an infinitesimal segment of a string in Fig. 3.1.

At the left end at $x$, the neighboring segment exerts vertical force in proportion to the vertical component of the tension $T(x)$. Similarly, at the right end at $x + dx$, the neighboring segment exerts vertical force in proportion to the vertical component of the tension $T(x + dx)$. Assuming that the tension along the string is constant, $T(x) = T(x + dx)$, and that the string makes angle $\theta_1$ and $\theta_2$ to a horizontal line at the left and right end of the segment, respectively, the vertical force acting on this segment at each end is given as follows.

© The Author(s), under exclusive license to Springer Nature Switzerland AG 2025    33
S. Yoshida, *Physics and Mathematics Behind Wave Dynamics*, Synthesis Lectures
on Wave Phenomena in the Physical Sciences,
https://doi.org/10.1007/978-3-031-60354-9_3

**Fig. 3.1** Force acting on two
ends of a string segment

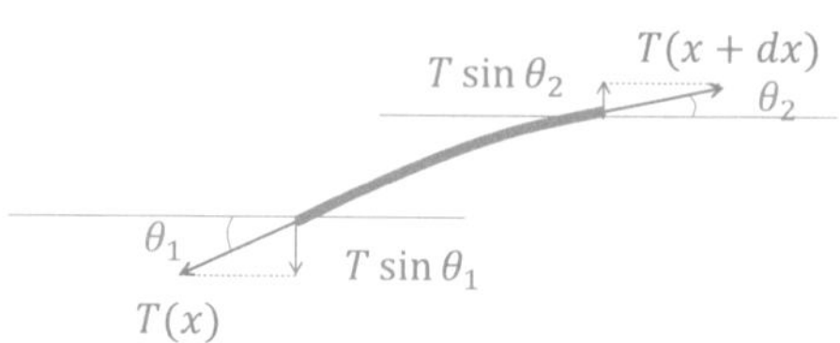



$$F_{left} = F_y(x) = T \sin \theta_1 \tag{3.1}$$

$$F_{right} = F_y(x + dx) = T \sin \theta_2 \tag{3.2}$$

From geometrical consideration and the assumption that angle $\theta_1$ and $\theta_2$ are small, we find
the following relations.

$$\sin \theta_1 \approx \tan \theta_1 = \left.\frac{d\xi_y}{dx}\right|_x \tag{3.3}$$

$$\sin \theta_2 \approx \tan \theta_2 = \left.\frac{d\xi_y}{dx}\right|_{x+dx} \tag{3.4}$$

From (3.1) through (3.4), we find the following expression for the net vertical force acting
on the segment.

$$F_{net} = F(x + dx) - F(x) = T \left( \left.\frac{d\xi_y}{dx}\right|_{x+dx} - \left.\frac{d\xi_y}{dx}\right|_x \right) = T \frac{d^2\xi_y}{dx^2} dx \tag{3.5}$$

Equation (3.5) leads to the following equation of motion.

$$\frac{m}{l} dx \frac{d^2\xi_y}{dt^2} = T \frac{d^2\xi_y}{dx^2} dx \tag{3.6}$$

Here $m$ and $l$ are the mass and length of the string. Using the linear density $\mu = m/l$, we
can rewrite (3.6) as follows.

$$\frac{d^2\xi_y}{dt^2} = \frac{T}{\mu} \frac{d^2\xi_y}{dx^2} \tag{3.7}$$

Equation (3.7) can be viewed as a wave equation that describes the vertical motion (displacement) of the string that travels along the string ($x$-axis) as a transverse wave. Here the
phase velocity of the wave appears on the right-hand side of the equation.

$$v_p = \sqrt{\frac{T}{\mu}} \tag{3.8}$$

Note that the wave velocity increases as the linear density decreases. This is intuitively
understood by noting that the heavier the string the lower the acceleration hence the longer
time is required for the string to oscillate.

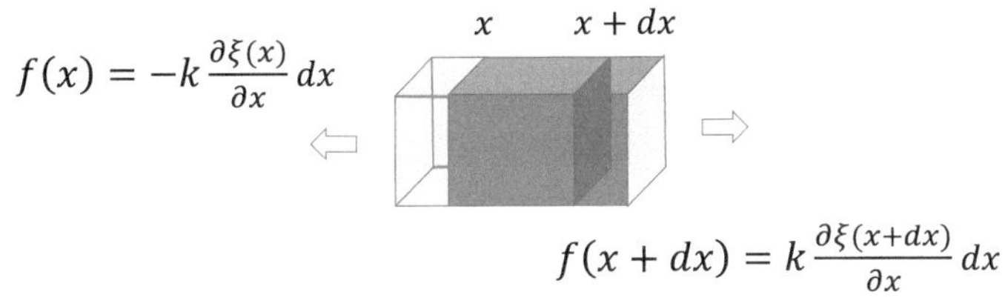

$$f(x) = -k\frac{\partial \xi(x)}{\partial x}\,dx$$

$$f(x+dx) = k\frac{\partial \xi(x+dx)}{\partial x}\,dx$$

**Fig. 3.2** An infinitesimally small (thin) block in an elastic object

### 3.1.2 Elastic Wave

A wave equation can be viewed as a special case of the equation of motion discussed in the preceding chapter. Consider an infinitesimally small block in an elastic object in Fig. 3.2. On the face of the block at $x$, the neighboring block on the left exerts elastic force $\mathbf{f}(x)$ in the negative $x$ direction. This elastic force is proportional to the stretch at $x$. Since stretch in this context is the differential displacement at $x$, the elastic force can be expressed as follows.

$$f_x(x) = -k_x\frac{\partial \xi_x(x)}{\partial x}\,dx \tag{3.9}$$

Here $f_x$ is the $x$-component of force vector $\mathbf{f}(x)$, $k_x$ is the spring constant (stiffness) of this elastic material in the $x$ direction, and $\xi_x$ is the $x$-component of the displacement vector.

Consider rewriting the stiffness using Young's modulus. Generally, the force $f_x$ can be expressed with the use of the spring constant and the normal strain as follows.

$$f_x = -k_x d\xi_x(x) = -k_x\frac{\partial \xi_x}{\partial x}\,dx = -AE\frac{\partial \xi_x}{\partial x} \tag{3.10}$$

Here $A$ is the cross-sectional area and $E$ is Young's modulus of this elastic material.

From (3.10), we find the following relation between Young's modulus and the spring constant.

$$k_x dx = AE \tag{3.11}$$

With the use of (3.11), (3.9) becomes as follows.

$$f_x(x) = -AE\frac{\partial \xi_x(x)}{\partial x} \tag{3.12}$$

On the other face at $x + dx$, the neighboring block on this side exerts an external force $f(x+dx)$ in the positive $x$ direction. Similarly to (3.12), the $x$-component of the elastic force here can be expressed as follows.

$$f_x(x+dx) = AE\frac{\partial \xi_x(x+dx)}{\partial x} \tag{3.13}$$

The net external force acting on this block in the $x$-direction is the differential force $df_x$. From (3.12) and (3.13), $df_x$ can be expressed as follows.

$$df_x = \frac{\partial f_x}{\partial x}dx = \frac{\partial}{\partial x}\left(AE\frac{\partial \xi_x(x+dx)}{\partial x} - AE\frac{\partial \xi_x(x)}{\partial x}\right)dx$$

$$= AE\frac{\partial}{\partial x}\frac{\partial \xi_x(x)}{\partial x}dx \tag{3.14}$$

Thus, the equation of motion associated with the dynamics of this block in the $x$-direction can be given as follows.

$$\rho A dx \frac{\partial^2 \xi_x}{\partial t^2} = df_x = AE\frac{\partial^2 \xi_x}{\partial x^2}dx \tag{3.15}$$

Canceling the common factors from the leftmost and rightmost terms of (3.15), we can express the wave equation in the following form.

$$\frac{\partial^2 \xi_x}{\partial t^2} = \frac{E}{\rho}\frac{\partial^2 \xi_x}{\partial x^2} \tag{3.16}$$

Since the direction in which the oscillatory motion is transferred and the displacement of the displacement are in line with each other, (3.16) is the wave equation of a longitudinal wave propagating through the elastic object. It is easily proved that the wave equation for a transverse wave can be expressed as follows [1].

$$\frac{\partial^2 \xi_x}{\partial t^2} = \frac{G}{\rho}\frac{\partial^2 \xi_x}{\partial x^2} \tag{3.17}$$

Here $G$ is the shear modulus of the elastic material.

More generally, the wave equation can be put in the following form.

$$\frac{\partial^2 \xi}{\partial t^2} = \frac{K}{\rho}\frac{\partial^2 \xi}{\partial x^2} \tag{3.18}$$

Here $K$ represents an elastic modulus and $\xi$ the displacement.

Equation (3.18) represents a wave propagating along the $x$-axis. We can extend the expression for a wave propagating in a general direction in the $xyz$-coordinate space by replacing the spatial differentiation $\partial^2/\partial x^2$ with Laplacian $\nabla^2$.

$$\frac{\partial^2 \xi}{\partial t^2} = \frac{K}{\rho}\nabla^2 \xi \tag{3.19}$$

In this case, since the wave equation takes into account the spatial variation three-dimensionally, it is often called the three-dimensional wave equation.[1] For the derivation of (3.19), see Ref. [1].

---

[1] This naming is somewhat misleading because it implies the wave propagates freely in the three-dimensional space. The wave still propagates in a certain direction.

### 3.1.3  Phase Velocity

The factor appearing on the right-hand side of the wave equation (3.18) represents the ratio of the temporal variation of the wave function to the spatial variation. If the value of this quantity is high, the temporal frequency is greater than the spatial frequency. This means that the wave oscillates faster in time and its spatial period (the reciprocal of the spatial frequency or the wavelength) is longer. Consequently, the spatial pattern travels faster. The square root of this quantity is known as the phase velocity $v_p$ of the wave.

$$v_p = \sqrt{\frac{K}{\rho}} \tag{3.20}$$

As (3.20) indicates, the phase velocity associated with a wave equation in the form of (3.19) is made up of the elastic constant and the density. These are material constants. This means that the phase velocity is determined by the material's properties. In other words, while a wave is excited by a certain source at a certain frequency, the resultant oscillation travels through the material at the velocity that the material chooses.

The phase velocity is not always constant. In some circumstances, it can depend on the frequency, or equivalently speaking, on the wave number. The phenomenon in which the phase velocity is a function of frequency is known as dispersion, and will be discussed in a later section (Sect. 4.5).

It is interesting and instructive to look at the phase velocity expression (3.20) in connection with the concept of natural frequency. Rewrite expression (3.20) for the longitudinal wave case using (3.11) as follows.

$$v_p = \sqrt{\frac{E}{\rho}} = \sqrt{\frac{E\,A\,dx}{\rho\,A\,dx}} = \sqrt{\frac{k_x\,dx^2}{\rho\,A\,dx}} = \sqrt{\frac{k_x}{m}}\,dx \tag{3.21}$$

Here $E$ is Young's modulus, $\rho$ is the density, $A$ is the cross-sectional area, $k_x$ is the spring constant, $dx$ is the infinitesimal length of the segment of interest, and $m$ is the mass of the segment. As we discussed n Sect. 2.2.1 (see (2.11)), $\sqrt{k_x/m}$ represents the natural (angular) frequency $\omega_0$. So, we can rewrite (3.21) as follows.

$$v_p = \omega_0 dx \tag{3.22}$$

We can view (3.22) as "the wave velocity per unit length is the natural frequency of the unit length". In other words, when a wave travels through an elastic medium due to the elastic recovery force, it oscillates the medium for each unit length at its natural frequency. This is how the elastic medium determines the phase velocity of a longitudinal wave. Figure 3.3 illustrates this view schematically. The same argument holds for a shear wave.

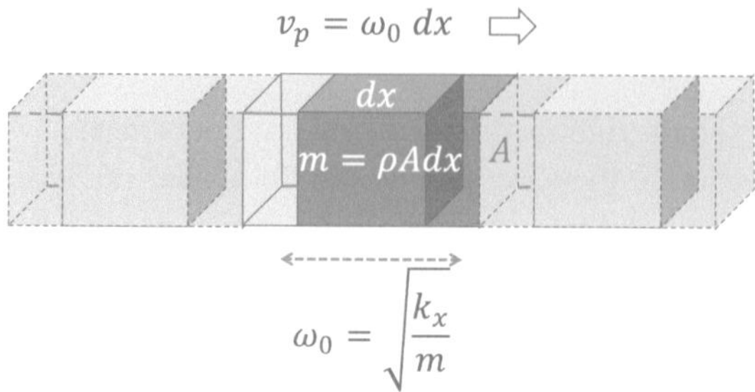



**Fig. 3.3** Schematic illustration of phase velocity as resonant oscillation per unit length. Each segment of length $dx$ and cross-sectional area $A$ oscillates at natural frequency $\omega_0 = \sqrt{k_x/m} = \sqrt{k_x/(\rho A dx)}$. This oscillation travels at phase velocity $v_p = \omega_0 dx$

### 3.1.4  Decaying Elastic Wave

We derived the wave equation (3.16) from the equation of motion (3.15), which takes into account only the elastic force as the external force acting on the block of the elastic object. As we will discuss in the following section, this type of wave equation yields a wave that travels forever. This is unrealistic. In reality, waves decay after traveling a certain distance. In Sect. 2.3, we introduced a velocity-damping mechanism in the equation of motion and discussed that the resultant solution represents a decaying oscillation. It is naturally understood that we can extend the argument made for an elastic wave to a decaying elastic wave by adding a velocity-damping term in the equation.

Thus, add the velocity damping force term to the wave equation (3.15) as follows.

$$\rho A dx \frac{\partial^2 \xi_x}{\partial t^2} = AE \frac{\partial^2 \xi_x}{\partial x^2} dx - b \frac{\partial \xi_x}{\partial t} \tag{3.23}$$

Dividing each term by $m = \rho A dx$ and using (2.31), rearrange (3.23) as follows.

$$\frac{\partial^2 \xi_x}{\partial t^2} + 2\beta \frac{\partial \xi_x}{\partial t} = \frac{E}{\rho} \frac{\partial^2 \xi_x}{\partial x^2} \tag{3.24}$$

For general cases where a decaying wave travels in a general direction, we can rewrite (3.24) in a similar form to (3.19).

$$\frac{\partial^2 \xi}{\partial t^2} + 2\beta \frac{\partial \xi}{\partial t} = \frac{K}{\rho} \nabla^2 \xi \tag{3.25}$$

Notice that when a wave experiences the damping effect, we cannot write the wave equation in the form of "temporal variation = constant $\times$ spatial variation", unlike the case of a non-decaying elastic wave. As discussed above, this is a case in which the wave experiences dispersion. More will be discussed in later sections.

## 3.2    Wave Solutions

An actual expression for a wave is given as a solution to the corresponding wave equation. For simplicity, let's discuss a solution to the wave equation (3.24) that represents a longitudinal decaying wave traveling along the $x$-axis. The discussions made here apply to other cases discussed in the previous section in general.

First, rearrange (3.24) into the following form for better mathematical visibility.

$$\frac{\partial^2 \xi_x}{\partial t^2} - \frac{E}{\rho}\frac{\partial^2 \xi_x}{\partial x^2} + 2\beta\frac{\partial \xi_x}{\partial t} = 0 \qquad (3.26)$$

In this form, we can interpret the wave equation as a differential equation where the right-hand side of the equation is a source term. Having a null source term, this differential equation describes the dynamics in which the elasticity maintains the oscillatory behavior of the medium propagates as a wave. The wave decays due to the damping property of the elastic medium, or in a more general sense, the property of the space through which the wave travels. The wave is generated at a certain point on the $x$-axis at a certain time by an external agent such as an acoustic transducer.

Let's solve the differential equation using a test function. We first find a plane wave solution. A plane wave has a plane wavefront, which means that the amplitude of the wave extends over an infinite area on the plane perpendicular to the axis of the propagation. As will be discussed shortly, this leads to a situation where the energy carried by the wave is infinite. Obviously, this is unrealistic. However, on many occasions, a plane wave solution is a good approximation, and more importantly, the argument made here for a plane wave solution contains important concepts that help us understand some basic behaviors of waves applicable to general cases.

We all know that a wave has temporal and spatial periodicity. The former represents the oscillation period and the latter is the wavelength. Mathematically, it is more convenient to use the reciprocals of these periods, frequencies, multiplied by $2\pi$ called the angular frequencies $\omega$ and $k$.[2] So let's put the test function in the following form.

$$\xi_x(t, x) = \xi_0 f(\omega t \pm kx) \qquad (3.27)$$

Here $\xi_0$ is the planar amplitude of the wave solution and $\xi_x(t, x)$ is a periodic function $f(t, x)$. When $t$ varies by an integer multiple of the period $\tau = 2\pi/\omega$ or $x$ varies by an integer multiple of the wavelength $\lambda = 2\pi/k$, the function takes the same value. The quantity inside the parenthesis determines the value between the maximum and minimum of the periodic function and is referred to as the phase of the wave.

The spatial frequency $k$ represents the rate at which the phase changes when the wave propagates along the $x$ axis. When a wave of this form propagates in a three-dimensional

---

[2] This $k$ is different from the spring constant $k_x$ used above.

direction represented by $\mathbf{r} = x\hat{x} + y\hat{y} + z\hat{z}$, the spatial part of the phase becomes $\mathbf{k} \cdot \mathbf{r}$ and the wave function takes the following form.

$$\xi_x(t, x) = \xi_0 f(\omega t \pm \mathbf{k} \cdot \mathbf{r}) \tag{3.28}$$

In this case, the wave propagates in the direction of $\pm\mathbf{r}$ at the rate $|k|$. From this viewpoint, $\mathbf{k}$ is called the propagation vector.

Substitution of test function (3.27) into differential equation (3.26) yields the following equation.

$$\omega^2 f'' - k^2 \frac{E}{\rho} f'' + \omega 2\beta f' = 0 \tag{3.29}$$

Here the prime $'$ represents the derivative of the function $f(t, x)$.

At a glance, we notice that differential equation (3.26) has the form of "summation of the temporal and spatial derivative is null". This indicates that the wave function, the solution, has the same form if the function is differentiated with respect to time or space, including the second-order spatial differentiation. We know that the exponential function has the property that it does not change the form when differentiated. In addition, from the above argument, the test function must be periodic. We know that when the exponent is a real number, the exponential function is not periodic. However, if the exponent is an imaginary number, the exponential function is periodic (Euler's formula). So, we rewrite expression (3.27) as follows.

$$\xi_x(t, x) = \xi_0 e^{i(\omega t \pm kx)} \tag{3.30}$$

Substituting (3.30) into (3.26) and dividing by the common term $exp\{i(\omega t \pm kx)\}$ in the resultant equation, we obtain the following equation known as the characteristic equation of differential equation.

$$\omega^2 - i2\beta\omega + k^2 \frac{E}{\rho} = 0 \tag{3.31}$$

The characteristic equation (3.31) does not hold when $\beta \neq 0$, and $\omega$ and $k$ are real, because the first and third terms are real and the second term is imaginary. It holds only when $\beta = 0$, or the wave does not decay. We can make the characteristic equation (3.31) hold for $\beta \neq 0$ cases by adding an imaginary term in the spatial angular frequency $k$ as $k = k_r - ik_i$. With this modification, (3.31) takes the following form.

$$\omega^2 - i2\beta\omega - (k_r^2 - k_i^2 - i2k_r k_i)\frac{E}{\rho} = 0 \tag{3.32}$$

Setting the real and imaginary parts to zero respectively, we obtain the following pair of equations.

$$\omega^2 = \frac{E}{\rho}\left(k_r^2 - k_i^2\right) \tag{3.33}$$

$$\beta\omega = \frac{E}{\rho}k_r k_i \tag{3.34}$$

Elimination of $k_i$ from (3.33) with the use of (3.34) yields the following equation.

$$\frac{E}{\rho} k_r^4 - \omega^2 k_r^2 - (\beta\omega)^2 \frac{\rho}{E} = 0 \tag{3.35}$$

By viewing (3.35) as a quadratic equation of $k_r^2$, we can solve it for $k_r^2$ and find $k_r$ as a function of $\omega$, $\rho$, $E$ and $\beta$. Once we find $k_r$, we can find $k_i$ from (3.33). The resultant expressions of $k_r$ and $k_i$ are as follows.

$$k_r = \omega \sqrt{\frac{\rho}{2E}} \left[ 1 + \sqrt{1 + \left(\frac{2\beta}{\omega}\right)^2} \right]^{\frac{1}{2}} \tag{3.36}$$

$$k_i = \omega \sqrt{\frac{\rho}{2E}} \left[ -1 + \sqrt{1 + \left(\frac{2\beta}{\omega}\right)^2} \right]^{\frac{1}{2}} \tag{3.37}$$

Equations (3.36) and (3.37) are the necessary conditions for the test function of the following form to be a solution to the differential equation.

$$\xi_x(t, x) = \xi_0 e^{-k_i x} e^{i(\omega t \pm k_r x)} \tag{3.38}$$

We can interpret that the real part $k_r$ represents the spatial periodicity of the wave and the imaginary part the decaying feature of the wave corresponding to the distance known as the penetration depth.[3] From this perspective, $\omega/k_r$ can be interpreted as the phase velocity of this decaying wave $v_p$.

$$v_p = \frac{\omega}{k_r} = \sqrt{\frac{2E}{\rho}} \left[ 1 + \sqrt{1 + \left(\frac{2\beta}{\omega}\right)^2} \right]^{-\frac{1}{2}} \tag{3.39}$$

Notice that the phase velocity in this case depends on the frequency $\omega$. Thus, the decaying wave undergoes dispersion. When the elastic medium does not exert the velocity damping force, $\beta = 0$. In this case, as expected, the phase velocity expression (3.39) reduces to the phase velocity expression for the non-decaying wave case (3.21). When $\omega >> \beta$, the dispersion is negligible. This situation can be interpreted as follows. Since the frequency is significantly higher than the decay rate, the wave undergoes a great number of oscillations before it decays. Consequently, the wave travels as if there is no damping effect, hence its velocity is close to the non-dispersive case.

---

[3] In this context, the penetration depth is defined as the distance where the amplitude of the wave decreases by a factor of $e$. Some authors define the penetration depth to be the distance where the intensity of the wave decreases by $e$. With this definition, $k_i$ corresponds to the point where the amplitude decreases by $\sqrt{e}$.

**Some Comments on Phase Velocity**

It is worthwhile taking some time to discuss the wave velocity a little deeper using the phase velocity expression (3.39). We can view the phase velocity from two perspectives. The first is the response of the medium[4] that the wave travels and the second is the rate that the phase information is transferred through the space.

View the phase velocity expression from the first perspective under the non-dispersive condition. Putting $\beta = 0$ leads to $v_p = \sqrt{E/\rho}$. As we discussed with (3.21), the elastic medium responds to the oscillation provided by the source of the wave by transferring the dynamics at the resonant frequency of the unit length. The greater the elastic constant, the faster the medium transfers the dynamics as the medium responds to the disturbance with a higher oscillation frequency. When the medium is subject to a damping mechanism, the rate of this transferring activity is altered through parameter $\beta$. Remember in Sect. 2.2 we discussed that the damping effect lowers the oscillation frequency from the natural frequency of the oscillatory system (the damping mechanism causes dispersion). In the present context, we can interpret that as much as the damping effect dissipates the oscillatory energy the medium becomes less responsive in the elastic recoverability, and that reduces the phase velocity.

The second perspective is independent of the medium's response. It is related to the relation between the temporal and spatial periodicity of the wave dynamics. Substitution of test function (3.30) into wave equation (3.18) with the use of (3.20) yields the following relation between $k$ and $\omega$.

$$v_p = \frac{\omega}{k} \tag{3.40}$$

We can view the right-hand side of (3.40) as the ratio of the temporal periodicity (frequency) to the spatial periodicity (frequency). The phase velocity indicates how quickly the oscillation (news) triggered by a source propagates through the medium. If the medium offers a high phase velocity, the news travels for a longer distance in a given time. From this standpoint, it makes sense to call the quantity $k$ the propagation constant.

If the medium is not dispersive, as we observed above, the propagation constant is a real number. Since the temporal frequency $\omega$ provided by the source is a real number, the phase velocity is a ratio of a real number to another real number. It is a real constant. If the medium is dispersive, the propagation constant is a complex number. Consequently, the ratio of the temporal frequency, which is a real number, to the propagation constant is not a deterministic material constant. It depends on the imaginary part of $k$, which represents the effect of velocity damping. As we observed in Sect. 2.4.2, the oscillation dynamics depends on the frequency when the spectrum is broadened due to the damping effect (if there is no damping, the spectrum does not depend on the frequency except at the resonant frequency). This naively explains the dispersion due to velocity damping.

---

[4] The word medium in this context includes vacuum.

### 3.2.1  Amplitude and Phase of Waves

Every wave carries energy. Consider the above-discussed longitudinal wave from this viewpoint.

By definition, energy is the scalar product of displacement and force. In the present context, since the wave is longitudinal, the displacement and the force are in line with each other. As the wave travels, at each transverse plane the displaced particle exerts force on the elastic material (the particle next to it). The energy passing through the transverse plane per unit time is the product of the particle velocity and the normal stress. With the displacement wave solution (3.30), the velocity $v_x(t, x)$ and the stress $\sigma_x(t, x)$ can be expressed as follows.

$$v_x(t, x) = \frac{\partial \xi_x}{\partial t} = i\omega\xi_0 e^{i(\omega t \pm kx)} \equiv v_0 e^{i(\omega t \pm k_r x)} \tag{3.41}$$

$$\sigma_x(t, x) = E\frac{\partial \xi_x}{\partial x} = i E k_x \xi_0 e^{i(\omega t \pm kx)} \equiv \sigma_0 e^{i(\omega t \pm k_r x)} \tag{3.42}$$

Here the $x$-dependent amplitude of the velocity and stress wave, $v_0$ and $\sigma_0$, are defined as follows.

$$v_0 \equiv i\omega\xi_0 e^{-k_i x} \tag{3.43}$$

$$\sigma_0 \equiv i E k_x \xi_0 e^{-k_i x} \tag{3.44}$$

Expressions (3.41) and (3.42) indicate that the velocity wave and stress wave have the common phase term $exp(i(\omega t \pm k_r x))$. Expressions (3.43) and (3.44) indicate that the velocity and stress waves have the same decay characteristics. It follows that the velocity and stress waves are in phase and decay in the same fashion along the axis of propagation.

The peak power can be evaluated as the product of $v_0$ and $\sigma_0$, and the root mean square (rms) power is $1/\sqrt{2}$ of the peak power.

$$P_{peak} = v_0\sigma_0 \tag{3.45}$$

$$P_{rms} = \frac{P_{peak}}{\sqrt{2}} = \frac{v_0\sigma_0}{\sqrt{2}} \tag{3.46}$$

It is instructive to find the unit of the power. Since $v_0$ is in m/s and $\sigma_0$ is in N/m$^2$, the unit of their product is $(N/m^2)\times(m/s)=J/(m^2 s)=W/m^2$. The power carried by the wave is the power density or the power per unit cross-sectional area. This quantity is referred to as the intensity of the wave.

Equations (3.43) and (3.44) indicate that the amplitudes of the stress and displacement are proportional to each other. The constant of the proportionality is called the impedance.

$$z_l = \frac{\sigma_0}{v_0} = \frac{E k_x}{\omega} = \frac{E}{v_p} = \sqrt{E\rho} \tag{3.47}$$

Here $z_l$ stands for the impedance for a longitudinal wave and $v_p$ is the phase velocity defined by (3.21). In the context of acoustic technology, we often refer to the quantity represented by $z_l$ as acoustic impedance.

Since the velocity and stress are in phase connected by a single quantity impedance, it is a convention to call the square of the amplitude of the velocity or stress wave the intensity of the wave. In other words, we can evaluate the product $v_0\sigma_0$ from $v_0^2$ or $\sigma_0^2$ by knowing the impedance of the medium.

Remember that a plane wave has a plane wavefront and therefore its amplitude extends over an infinite area on the plane perpendicular to the axis of the propagation. Now we just found that the square of the amplitude of a wave represents the power density. These mean that a plane wave has infinite power and that is not realistic. In the next section, we discuss a more realistic wave called the paraxial wave.

## 3.3 Paraxial Wave Approximation and Helmholtz Equation

A paraxial wave is a wave whose intensity is localized in the vicinity of the axis of propagation. What does the word "vicinity" mean? It is defined by the source of the wave. When a wave is generated with a source, the source has a finite dimension transverse to the axis of the wave propagation. It is impossible to generate a wave outside of this transverse dimension of the source. This defines the starting (the initial) transverse size of the wave. The vicinity is naturally defined by this initial transverse size of the wave.

When the wave source has a finite transverse size, the wave exhibits natural diffraction. This causes two main differences from the plane wave case. First, the radius of the wave (beam) increases as the wave propagates, Second, the wavefront is curved. The former is obvious from the definition of diffraction. The latter may not be as obvious but is intuitively understood by considering a little rock thrown into a calm pond. This small wave source will generate a circular wavefront that spreads concentrically.

Mathematically, a paraxial solution is given as a solution to a differential equation known as the Helmholtz equation. As will be clarified below, with certain approximations, the solution exhibits a transverse profile of the amplitude in the form of a Gaussian distribution. Thus, the solution is generally called a Gaussian beam. Since the approximations are used in the solution process, the resultant Gaussian beam is a paraxial approximation of the wave. In most applications, this approximation is accurate enough to describe the wave propagation and related phenomena such as the focusability of the wave and the focal size. A Gaussian beam solution tells us the beam size (the wave radius) and the wavefront (the radius of curvature) at a given point on the propagation axis.

Let's start with the three-dimensional wave equation (3.19). First, separate the solution function $\xi(t, x, y, z)$ into a time-dependent and space-independent function $T(t)$, and a time-independent and space-dependent function $S(x, y, z)$.

$$\xi(t, x, y, z) = \xi_0 T(t) S(r) \tag{3.48}$$

Here $r$ is the radial distance of a coordinate point from the origin; $r = \sqrt{x^2 + y^2 + z^2}$. Substitution of (3.48) into differential equation (3.19) and rearrangement of terms yield the following equation.

$$\frac{\ddot{T}}{T} = v_p^2 \frac{\nabla^2 S}{S} = -\omega^2 \tag{3.49}$$

Here $K/\rho$ in (3.19) has been replaced with the square of the phase velocity $v_p$ defined by (3.20). The rightmost-hand side of (3.49) comes from the assumption that the oscillation is in the following form.

$$T(t) = T_0 e^{i\omega t} \tag{3.50}$$

Using expression (3.40) for $v_p$, we can rewrite (3.49) in the following form, which is known as the Helmholtz equation.

$$\nabla^2 S + k^2 S = 0 \tag{3.51}$$

Here we can identify $k$ as the magnitude of the propagation vector defined by (3.28) by assuming that $S(r)$ has a sinusoidal variation spatially. Set the $z$-axis in the direction of the propagation. In this case, we can put $S(r)$ in the following form.

$$S(r) = \Psi(x, y, z) e^{-ikz} \tag{3.52}$$

Here $k$ is a constant because we assume that the wave propagates through a uniform medium. If the medium is not uniform, $k$ becomes a spatial function. When a laser beam travels through a medium experiencing the thermal lensing effect [2], for example, the index of refraction can have a quadratic dependence in the transverse direction. In that case $k$ is a function of $x$ and $y$; $k(x, y)$.

After somewhat lengthy mathematical manipulation, we find the following form of $\Psi$ satisfies (3.49).

$$\Psi = \Psi_0 e^{-i\left[P(z) + \frac{1}{2} Q(z) r^2\right]} \tag{3.53}$$

where

$$e^{-iP(z)} = \frac{1}{\sqrt{1 + \left(\frac{z}{z_0}\right)^2}} e^{i \tan^{-1}\left(\frac{z}{z_0}\right)} \tag{3.54}$$

$$e^{-i\frac{1}{2} Q(z) r^2} = e^{-i\frac{k r^2}{2(z + i z_0)}} \tag{3.55}$$

By defining the following quantities,

$$R(z) = z\left(1 + \left(\frac{z_0}{z}\right)^2\right) \tag{3.56}$$

$$w_0 = \sqrt{\frac{2 z_0}{k}} = \sqrt{\frac{\lambda z_0}{\pi}} \tag{3.57}$$

$$w(z) = w_0 \sqrt{1 + \left(\frac{z}{z_0}\right)^2} \tag{3.58}$$

we can simplify expressions (3.54) and (3.55), and rewrite expression (3.53) in the following form.

$$\Psi(r, z) = \Psi_0 \frac{w_0}{w(z)} e^{-\frac{r^2}{w(z)^2}} e^{i\tan^{-1}\left(\frac{z}{z_0}\right)} e^{-i\frac{kr^2}{2R(z)}} \tag{3.59}$$

Here, $R(z)$ is the radius of curvature of the wavefront (the curved plane of a constant phase) at $z$, parameter $z_0$ is known as Rayleigh length (of which physical meaning will be discussed shortly), and $w(z)$ is the radial distance from the $z$-axis where the amplitude of $\Psi$ falls off to $1/e$ of the axial value (see the first term of the exponential function in (3.59)). $w(z)$ is called the spot size of the Gaussian beam at $z$. Rayleigh length is defined with wavelength $\lambda$ and beam waist as follows.

$$z_0 = \frac{\pi w_0^2}{\lambda} \tag{3.60}$$

Equation (3.58) indicates that $w_0$ is the minimum spot size, and is called the beam waist. Figure 3.4a illustrates $w(z)$ as a function of radial distance from the beam axis. As (3.59) indicates, $\Psi(r, z)$ represents a Gaussian distribution of the wave's amplitude at a given axial position $z$. For $z = 0$ m the spot size is indicated. Figure 3.4b illustrates that the spot size $w(z)$ increases in going either direction along the $z$-axis from $z = 0$ (see (3.58)). That is why $w(0) = w_0$ is referred to as the beam waist. The spot size indicated in Fig. 3.4a is the beam waist size as this Gaussian beam has the beam waist at the axial position $z = 0$ m. Figure 3.4c shows that the radius of curvature $R(z)$ takes the minimum value at the Rayleigh length and keeps increasing from there. The asymptotic behavior of $R(z)$ will be discussed shortly.

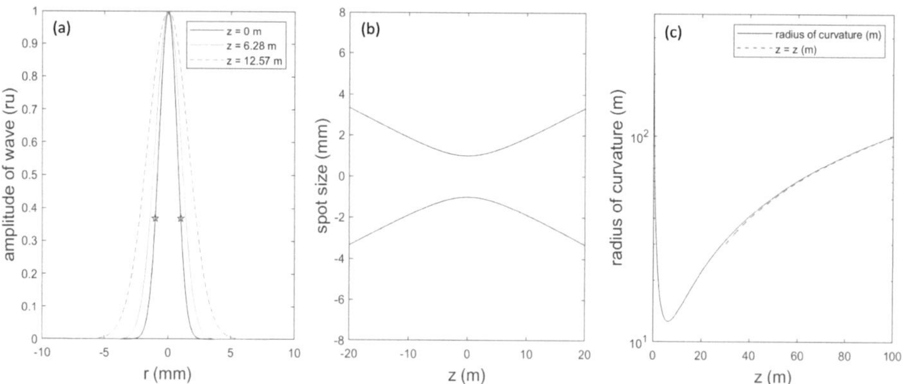


**Fig. 3.4** Gaussian beam **a** Gaussian amplitude profile at three axial locations ($z$). Spot size (beam waist size) is indicated for $z = 0$ m. **b** Beam spot size variation along beam axis; **c** Radius of curvature variation

With all the above-defined expressions, the entire space part of the solution (3.52) takes the following form.

$$S(r, z) = \Psi_0 \frac{w_0}{w(z)} e^{-\frac{r^2}{w(z)^2}} e^{-i \frac{kr^2}{2R(z)}} e^{-i\left[kz - \tan^{-1}\left(\frac{z}{z_0}\right)\right]} \tag{3.61}$$

The wave expressed in the form of expression (3.61) is called a Gaussian beam. On the right-hand side of (3.61), the physical meaning of each term is as follows.

**On-Axis Amplitude Variation $\Psi_0 \frac{w_0}{w(z)}$**

This term represents the variation of amplitude as a function of $z$. Notice that at the beam waist $(w(0) = w_0)$, the fraction $w_0/w(z)$ takes the maximum value of unity and therefore at this location the amplitude is equal to $\Psi_0$. As the beam propagates along the $z$-axis, in the positive or negative direction, the value of $w(z)$ increases, and thereby the fraction $w_0/w(z)$ decreases. This indicates that as the wave diverges, its intensity (the power per unit cross-sectional area) decreases because the total power is constant.

**Transverse Amplitude (Intensity) Profile $e^{-\frac{r^2}{w(z)^2}}$**

This term indicates that (a) at a given on-axis location $z$ the transverse intensity distribution of the Gaussian beam is Gaussian and (b) as the beam travels further away from the beam waist the Gaussian profile spreads out transversely (because $w(z)$ increases). The beam spreads out because of diffraction.

**Transverse Phase Variation $e^{-i \frac{kr^2}{2R(z)}}$**

This term indicates that (a) at a given $z$ location, the transverse phase profile is quadratic $e^{-i \frac{kr^2}{2R(z)}}$ and (b) the curvature of the quadratic function varies with the variation of $R(z)$ as a function of $z$. Figure 3.5 plots the phase of the Gaussian beam at $z = z_0$ and $z = 2z_0$ as a function of radial distance from the beam axis. The parabolic dependence is clearly seen. It follows from (a) that the wavefront, the two-dimensional surface of a constant phase, is parabolic, and from (b) that the curvature of the parabolic surface is $R(z)$. Since $R(z)$ takes the minimum value at $z = z_0$, the parabolic surface has the largest curvature at the Rayleigh length from the beam waist position. (3.56) indicates that after the Rayleigh length, the radius of curvature increases approaching the line $R = z$. This indicates that at a distance significantly greater than the Rayleigh length the radius of curvature is approximately equal to the distance from the beam waist location. The dashed line inserted in Fig. 3.4c shows this asymptotic behavior of $R(z)$.

**On-Axis Phase Variation $e^{-i\left[kz - \tan^{-1}\left(\frac{z}{z_0}\right)\right]}$**

This term represents the on-axis phase variation. Of the two terms in the exponent, the first term represents the change in the phase from the beam waist location in the unit of the wave number; this is the same as the plane wave case. The second term is the additional factor associated with the non-planar nature of the wavefront. The shorter the Rayleigh length, the greater this effect as the fraction $z/z_0$ becomes greater. In the plane wave limit where $z_0 \rightarrow \infty$ this term approaches zero.

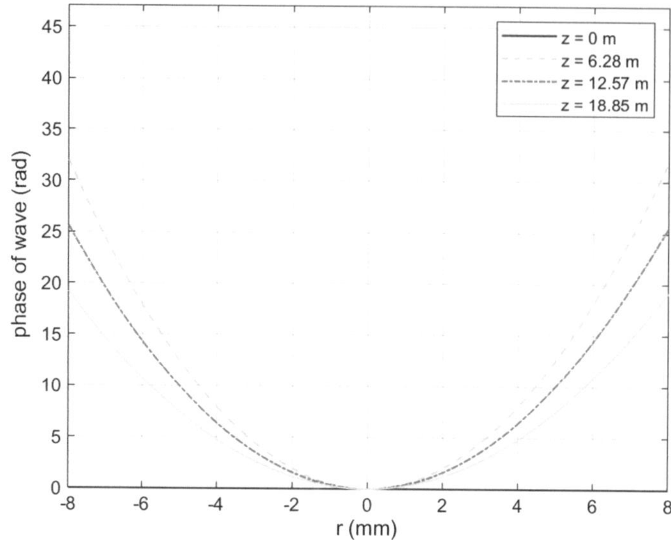

**Fig. 3.5** Gaussian beam phase as a function of radial distance at several locations on the beam axis

### 3.3.1  Propagation of Gaussian Beam

Gaussian beams are useful in various applications of acoustics and optics. Here, we consider some basic properties of the propagation of a Gaussian beam.

One important feature of the Gaussian beam is that once the beam waist size (the value of $w_0$) and its location (the value of $z$ coordinate point where $w(z) = w_0$) are given, the Gaussian beam is uniquely specified. In other words, a set of beam waist size and location determines the beam size and radius of curvature for the entire space, i.e. $w(z)$ and $R(z)$ are given as a function of $z$ at any point on the $z$-axis. Figure 3.6 illustrates the variation of beam size and radius of curvature for two Gaussian beams propagating in a free space with two different waist sizes located at the same location on the beam axis. Note that the one with the smaller beam waist size expands faster than the larger one along the $z$-axis. Figure 3.6b indicates that the smaller beam waist one has the shorter minimum radius of curvature at the Rayleigh length.

In practical applications, it is often the case that a Gaussian beam passes through objects that change the wavefront curvature. An acoustic or optical lens is a good example of such an object. The wavefront curvature also changes when a wave passes through a boundary of media having different refractive properties, hence the phase velocity changes.

We use a refractive index, $n$, to indicate the change in the phase velocity due to the refractive properties of a medium relative to a reference medium. In the case of acoustic waves, water is usually used as the reference while for optical waves vacuum is used as the reference. (The speed of light in air is approximately equal to vacuum. So, practically air

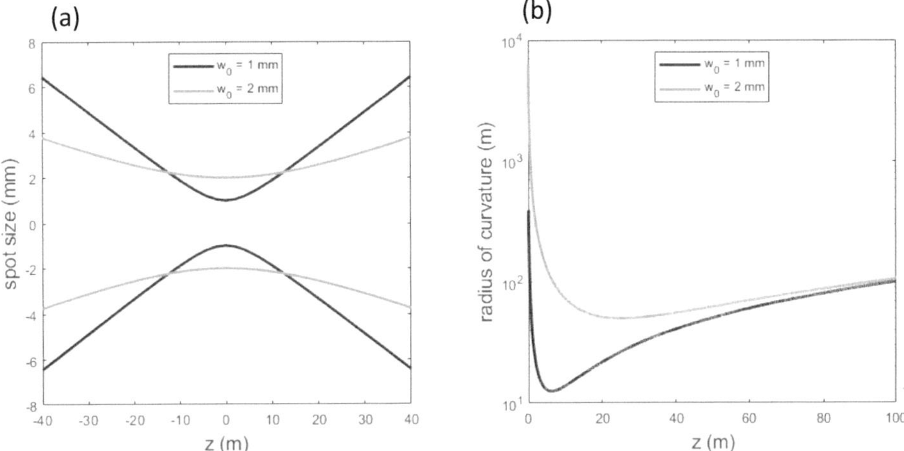

**Fig. 3.6 a** Spot size and **b** radius of curvature of Gaussian beams with two beam waist sizes

can be used as the reference for optical waves.) In general, the phase velocity of a wave is expressed as follows. (Sect. 4.1)

$$v_p = \frac{v_0}{n} \tag{3.62}$$

Here $v_p$ and $n$ are the phase velocity and refractive index of the medium of interest, and $v_0$ is the phase velocity in the reference medium. Note that the higher the refractive index the lower the wave's phase velocity.

Figure 3.7 illustrates the situation where a Gaussian beam passes through a boundary from the left to the right. As the beam crosses the boundary the wave velocity decreases from $v_1$ to $v_2 = v_1/1.5$. Figure 3.7a shows the spot size as a function of axial distance where the dashed line indicates how the spot size of the incident wave increases with z when the boundary does not exist. Figure 3.7b illustrates the behavior of the radius of curvature where the dashed line is the case when the boundary does not exist. It is seen that as the wave propagates more slowly, the beam divergence due to diffraction becomes slower resulting in the spot size increasing more slowly (Fig. 3.7a) and the radius of curvature increases more slowly than before (Fig. 3.7b). Naively speaking, we can understand this as "after passing through the boundary the diffraction proceeds more slowly because the wave velocity becomes lower. The wave enters a world where changes become slower".

Often we need to find the beam size ($w(z)$) and wavefront curvature ($R(z)$) of a Gaussian wave while it passes through multiple media having different refractive properties. Below we briefly discuss ways to find $w(z)$ and $R(z)$ for two cases. The first case is when a Gaussian beam propagates in a space at a constant wave velocity. The second case is when a Gaussian beam passes through objects having different refractive properties.

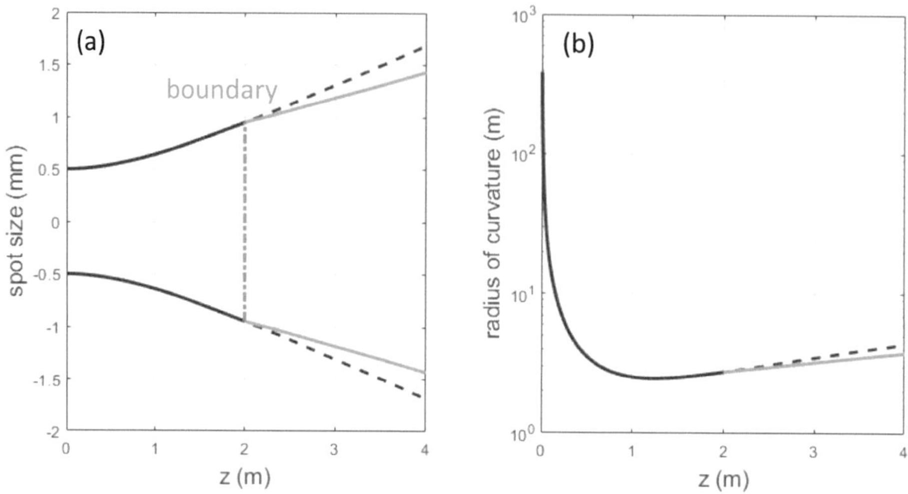


**Fig. 3.7** Gaussian beam passing through boundary of media with different refractive indices

### Case 1: When a Gaussian Beam Propagates Through a Space with Uniform Refractive Properties

In this case, we can simply use expressions (3.58) and (3.56) to find $w(z)$ and $R(z)$. In doing so, it is important to use the correct value of $z$. If the beam waist is not located at $z = 0$ it is necessary to consider the distance from the origin of the $z$-axis and the beam waist point (the offset). See Case 2 below for a convenient way to consider the offset.

### Case 2: When Objects Exist in the Beam Path

In this case, the quantity known as the complex radius $q(z)$ of the Gaussian beam is extremely useful. This quantity is defined as follows.

$$q(z) = \frac{k}{Q(z)} \tag{3.63}$$

where $k$ is the propagation constant and $Q(z)$ is defined by (3.53). At a given point $z$ where $w(z)$ and $R(z)$ are known, we can evaluate $q(z)$ as follows. (Use (3.55), (3.56), (3.57), and (3.58) to derive (3.64)).

$$\frac{1}{q(z)} = \frac{1}{R(z)} - i\frac{\lambda}{\pi w(z)^2} \tag{3.64}$$

Once $q(z)$ is evaluated, we can propagate the Gaussian beam to the next object that changes the curvature with the following equation.

$$q(z + \Delta z) = q(z) + \Delta z \tag{3.65}$$

Here $\Delta z$ is the distance from the current location where $q(z)$ is evaluated to the point where the next object is located. To know the beam size and curvature at $z + \Delta z$, we can take the following steps.

Step (1) Use (3.65) to evaluate the value of $q$ at $z + \Delta z$.

Step (2) Substitute the value of $q(z + \Delta z)$ evaluated in step (1) into the left-hand side of (3.64)

**Table 3.1** Ray matrix for typical optical objects. $n$ is the refractive index. The phase velocity of light is inversely proportional to the refractive index as $v_p = v_0/n$. Here $v_0$ is the speed of light in vacuum.

| Object | $\begin{pmatrix} A & B \\ C & D \end{pmatrix}$ | Beam propagation |
|---|---|---|
| Free space | $\begin{pmatrix} 1 & d \\ 0 & 1 \end{pmatrix}$ | 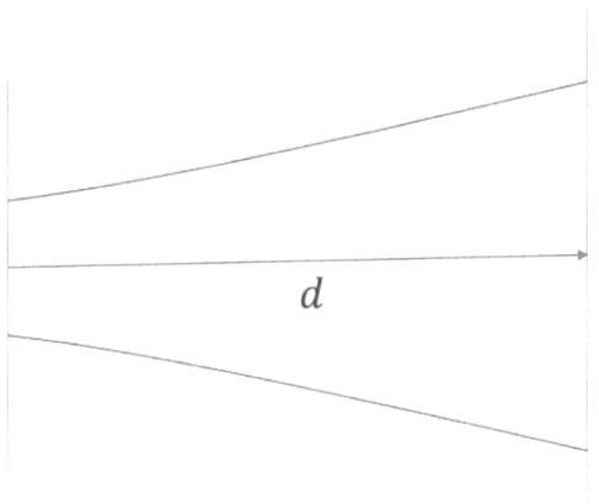 |
| Thin lens | $\begin{pmatrix} 1 & 0 \\ \frac{-1}{f} & 1 \end{pmatrix}$ | 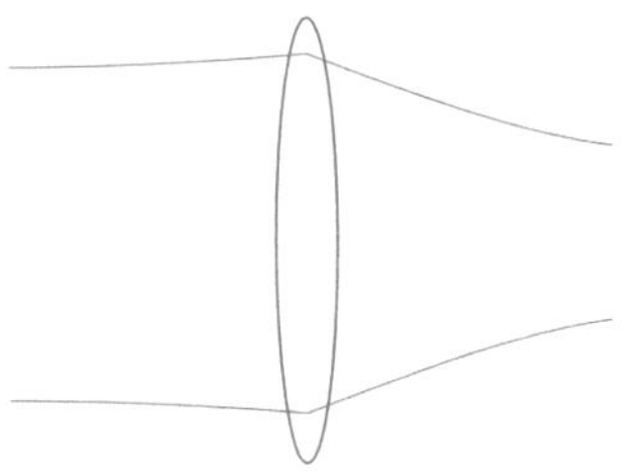 |
| Boundary $n_1$ to $n_2$ | $\begin{pmatrix} 1 & 0 \\ 0 & \frac{n_1}{n_2} \end{pmatrix}$ | 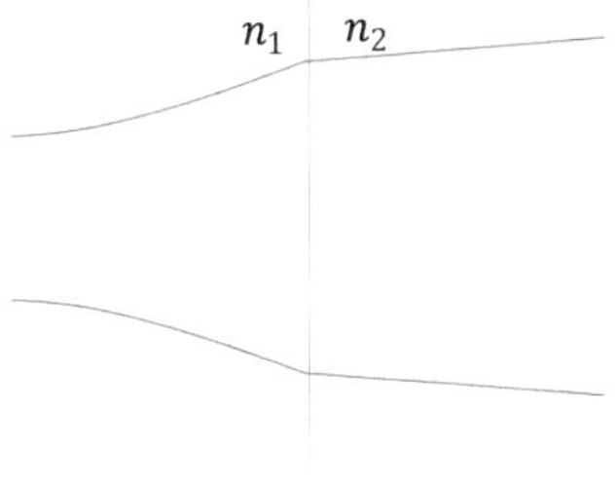 |

Step (3) Find $w(z + \Delta z)$ and $R(z + \Delta z)$ from the real and imaginary parts of the complex number $1/q(z + \Delta z)$

$$\frac{1}{q(z + \Delta z)} = \frac{1}{R(z + \Delta z)} - i\frac{\lambda}{\pi w(z + \Delta z)^2} \tag{3.66}$$

In this fashion, we can evaluate the spot size and radius of curvature at the location where the next object is placed.

Step (4) Find complex radius output from the object

The complex radius $q(z + \Delta z)$ evaluated in step (4) is the input to the object placed at $z + \Delta z$, called $q_{in}$. We can find the complex radius of the output beam from the object, called $q_{out}$ using the following formula [3].

$$q_{out} = \frac{Aq_{in} + B}{Cq_{in} + D} \tag{3.67}$$

Here each object has a unique expression for A, B, C, and D. These quantities A - D form a matrix known as theABCD or ray matrix [4]. Table 3.1 lists the ray matrix of typical optical objects, as an example.

Once we find $q_{out}$ we can repeat the above four steps to find the complex radius (and the spot size and radius of curvature as needed) of the output beam from the next object. In this fashion, we can evaluate the Gaussian beam's spot size and radius of curvature at any point on the $z$ (propagation) axis.

---

## References


1. Yoshida S (2017) Waves; Fundamental and dynamics. San Rafael, CA, USA, IOP Publishing, Bristol, UK, Morgan & Claypool
2. Mawatari K, Kitamori T (2024) Thermal lensing, detection, In: Li D (eds) Encyclopedia of microfluidics and nanofluidics. Springer, Boston, MA. https://doi.org/10.1007/978-0-387-48998-8_1556. (Accessed on February 5, 2024)
3. Pedrotti FL, Pedrotti LM, Pedrotti LS (2018) Introduction to optics 3rd edn. Cambridge University Press, pp 593–597
4. Yariv A (1989) Quantum electronics, 3d edn. Wiley, New York. Chap. 6
5. Hecht E (2002) Optics, 4th edn. Addison Wesley, San Francisco, CA, USA. Chap. 4
6. Kirchhoff G (1883) Zur Theorie der Lichtstrahlen. Ann d Physik 2(18):663–695
7. Park IK, Park TS, Miyasaka C (2012) Quantitative NDE of the nano-scaled thin film system using SAM. In: 2012 IEEE international ultrasonics symposium, Dresden, pp 1742–1745. https://doi.org/10.1109/ULTSYM.2012.0437

# Properties of Waves

This chapter describes some basic properties of waves. After briefly describing the laws of reflection and refraction, and discussing the interference of waves, we consider the diffraction and dispersion of waves in some detail.

## 4.1 Reflection and Refraction

When a wave is incident obliquely to a boundary plane, the reflected and transmitted waves propagate in certain directions. These directions are represented by the angle of reflection and refraction, which is defined as the angle made by the propagation vector and a line normal to the boundary plane. We can use the laws of reflection and refraction to find the angles of reflection and refraction.

Figure 4.1 illustrates the reflection and refraction of a wave. Consider that a wave of wavelength $\lambda$ is incident to a planar boundary. Let $\theta_i$, $\theta_r$, and $\theta_t$ be the angle of incidence, reflection, and refraction respectively. Express the phase velocity of the wave in the medium on the incident wave side of the boundary, designated by 1, and that in the medium on the refracted wave side of the boundary, designated by 2, as follows.

$$v_1 = \frac{v_0}{n_1} \tag{4.1}$$

$$v_2 = \frac{v_0}{n_2} \tag{4.2}$$

Here $v_0$ is the reference phase velocity, and $n_1$ and $n_2$ are the index of refraction for the respective media.

© The Author(s), under exclusive license to Springer Nature Switzerland AG 2025

S. Yoshida, *Physics and Mathematics Behind Wave Dynamics*, Synthesis Lectures on Wave Phenomena in the Physical Sciences, https://doi.org/10.1007/978-3-031-60354-9_4

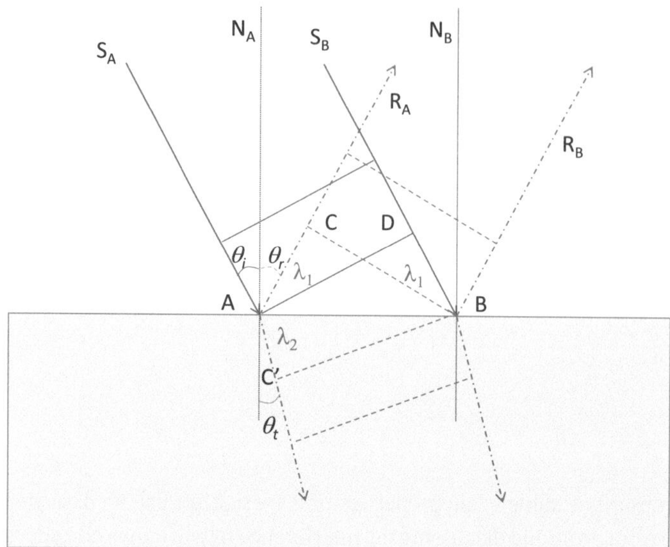



**Fig. 4.1** Reflection and transmission of wave

Since the frequency of the wave does not change at the boundary,[1] the wavelengths inside medium 1 and 2 become as follows.

$$\lambda_1 = \frac{\lambda_0}{n_1} \tag{4.3}$$

$$\lambda_2 = \frac{\lambda_0}{n_2} \tag{4.4}$$

Here $\lambda_0 = v_0/\nu$ with $\nu$ being the frequency.

With the above information, we can find the relationships between $\theta_i$ and $\theta_r$, and $\theta_i$ and $\theta_t$, respectively. The laws that describe these relationships are known as the law of reflection, and the law of refraction or Snell's law, respectively. In the following sections, we discuss these laws. There, we use a non-zero angle of incidence. When the angle of incidence is zero (referred to as the normal incidence), the reflected wave propagates in the opposite direction to the incident wave and the refracted (transmitted) wave propagates in the same direction as the incident wave.

Note that the laws of reflection and refraction describe the directions of the reflected and transmitted waves. The reflectance (the reflected portion of the wave energy) and transmittance (the transmitted portion of the wave energy) are determined by the boundary condi-

[1] The reason why the frequency remains the same through the boundary may not be straightforward. However, remember in Sect. 1.1 we characterized a wave as a moving oscillatory pattern. It can be naturally accepted that the frequency remains the same when an oscillation generated by a certain mechanism propagates as a wave.

tions. Since the boundary conditions depend on the type of the wave (e.g., optical waves, acoustic waves, etc.) we do not discuss reflectance or transmittance in this book. The readers interested in these topics are encouraged to refer to references, e.g., for optical waves see Refs. [1, 2] and acoustic waves see Ref. [3, 4].

### 4.1.1   Law of Reflection

First, consider the relation between $\theta_i$ and $\theta_r$. In Fig. 4.1, solid lines $S_A A$ and $S_B B$ are two representative propagation vectors of the incident wave and the dashed lines connecting these two lines are two representative wavefronts that are apart by one wavelength. Dash-dotted lines $AR_A$ and $BR_B$ are two representative propagation vectors of the reflected wave. The dashed lines associated with $AR_A$ and $BR_B$ are also wavefronts one wavelength apart. Compare triangles ADB and ACB, which share side AB. The angle made by lines AD and AB is equal to the angle of incidence because both are $90°$ minus angle $N_A AD$. Since the incident and reflected waves are in medium 1, their wavelengths are $\lambda_1$ and equal to each other. Therefore, these two triangles are congruent. Consequently,

$$\theta_i = \theta_r \tag{4.5}$$

Relation (4.5) is the law of reflection.

### 4.1.2   Law of Refraction

Next, consider the relation between $\theta_i$ and $\theta_t$ by comparing triangles ADB and AC'B. Here dash-dotted lines in medium 2 are two representative propagation vectors of the refracted wave and dashed lines associated with these propagation vectors are two wavefronts apart by one wavelength $\lambda_2$. In this case, triangles ADB and AC'B are not congruent because AC' $\neq$ BD. However, the following relation is true.

$$\lambda_1 = AB \sin \theta_i \tag{4.6}$$
$$\lambda_2 = AB \sin \theta_t \tag{4.7}$$

Substitution of (4.3) and (4.4) into (4.6) and (4.7) leads to the following equation.

$$n_1 \sin \theta_i = n_2 \sin \theta_t \tag{4.8}$$

Relation (4.8) is the law of refraction or Snell's law.

Equation (4.8) indicates that if a wave is incident to a medium whose index of refraction is greater ($n_1 < n_2$), the angle of refraction is smaller than the angle of incidence ($\theta_i > \theta_t$).

Reflection under this condition is referred to as external reflection. Conversely, when $n_1 > n_2$, it follows that $(\theta_i < \theta_t)$. Reflection associated with this condition is referred to as internal reflection.

Consider that the angle of incidence is increased when a wave is incident to a medium whose index of refraction is smaller. According to (4.8), the angle of refraction $\theta_t$ reaches $90°$ earlier than $\theta_i$. Call the angle of incidence that makes the angle of refraction equal to $90°$ the critical angle $\theta_c$. From (4.8),

$$\sin \theta_c = \frac{n_2}{n_1} \tag{4.9}$$

If the angle of incidence is greater than the critical angle $\theta_c$, it becomes impossible for the incident wave to pass through the boundary. Consequently, the incident wave is completely reflected. This phenomenon is referred to as total reflection. Total reflection occurs only in internal reflection, as it is always possible to have an angle of refraction less than $90°$ for external reflection.

## 4.2 Interference

When two or more waves share the same space at the same time, the physical quantities of individual waves (e.g., acoustic fields, electromagnetic fields, etc.) add to each other. Consequently, the shared space experiences the total field. As a superposition of waves, the resultant value of the total field depends on the relative phase. If the relative phase is such that the signs of the component waves are the same, the total amplitude is greater than that of a component wave. If the signs of the phase are opposite to each other, the total amplitude is lower than that of a component wave. This phenomenon is referred to as interference of waves.

The word "interfere" may sound like a negative effect but there are a number of important applications of interference for engineering and science. In short, interference is useful because it converts the phase information to intensity. Normally the direct measurement of a phase is more difficult than that of intensity. Through examination of interference patterns, we can retrieve the phase information. Often the phase is directly related to the path length of the wave, and from the phase information, we can perform length measurement. In some cases, from the measured length and the known wave velocity we can characterize the property of the medium. Such measurement techniques are generally called interferometry. Interferometry is essential to use waves as a ruler to conduct certain measurements.

Waves can interfere with one another regardless of the frequency of component waves. However, from the phenomenological and practical points of view, there is a significant difference between the cases where the waves being superposed have the same frequency and not. In the following sections, we discuss both cases separately.

## 4.2.1 Adding Two Waves of the Same Frequency

In this section, we discuss the addition of two waves that have a common frequency. The content of this section holds for the cases where more than two waves are superposed.

When waves having the same frequency are superposed, the relative phase difference is the most important factor in the intensity of the superposed wave. When waves interfere with one another, the interaction takes place at the same time and space. So, although the phases of waves vary as a function of time and space, we can use a common set of space-time coordinates to express the component waves. This also means that as long as the frequency is the same for the component waves, the phase difference remains the same.

Consider that the following two waves $f_1$ and $f_2$ are being superposed.

$$f_1 = A_1(x, y.z)\cos(\omega t - kl_1) \tag{4.10}$$

$$f_2 = A_2(x, y, z)\cos(\omega t - kl_2) \tag{4.11}$$

Here $A_1$ and $A_2$ are the amplitudes of the component waves at a coordinate point $P(x, y, z)$, and $l_1$ and $l_2$ are the distances to point $P(x, y, z)$ from the sources of the respective waves. For simplicity, use $\theta_1$ and $\theta_2$ to express the phase of the component waves, and rewrite expressions (4.10) and (4.11) using the Euler's notation. The intensity of the superposed wave can be expressed as follows. Below * represents the complex conjugate.

$$
\begin{aligned}
I_{total} =&(f_1 + f_2)(f_1 + f_2)^* \\
=&(A_1 e^{i\theta_1} + A_2 e^{i\theta_2})(A_1 e^{-i\theta_1} + A_2 e^{-i\theta_2}) \\
=&A_1^2 + A_2^2 + A_1 A_2(e^{i(\theta_1 - \theta_2)} + e^{-i(\theta_1 - \theta_2)}) \\
=&A_1^2 + A_2^2 + A_1 A_2\{\cos(\theta_1 - \theta_2) + i\sin(\theta_1 - \theta_2) \\
&\qquad\qquad\qquad + \cos(\theta_1 - \theta_2) - i\sin(\theta_1 - \theta_2)\} \\
=&A_1^2 + A_2^2 + 2A_1 A_2 \cos(\theta_1 - \theta_2) \tag{4.12}
\end{aligned}
$$

Equation (4.12) indicates that the total intensity depends on the amplitude and the relative phase of the component waves. In particular, when $A_1 = A_2 = A$ and $\theta_1 - \theta_2 = (2N + 1)\pi$, the total intensity becomes zero as follows. Here $N$ is an integer.

$$
\begin{aligned}
I_{total} &= A_1^2 + A_2^2 + 2A_1 A_2 \cos((2N + 1)\pi) = A_1^2 + A_2^2 - 2A_1 A_2 \\
&= (A_1 - A_2)^2 = (A - A)^2 = 0
\end{aligned} \tag{4.13}
$$

Figure 4.2 illustrates the total intensity as a function of the relative phase difference $\Delta\theta \equiv \theta_1 - \theta_2$. Here, Fig. 4.2a shows the amplitude of the first wave (wave 1) as a function of its phase, (b) shows the amplitude of the second wave (wave 2) for five-phase differences relative to wave 1, and (c) shows the intensity of the addition of two waves (total intensity) as a function of the relative phase difference. The total intensity for the five relative phase differences used in Fig. 4.2b are marked. It is seen that the intensity is at a maximum when

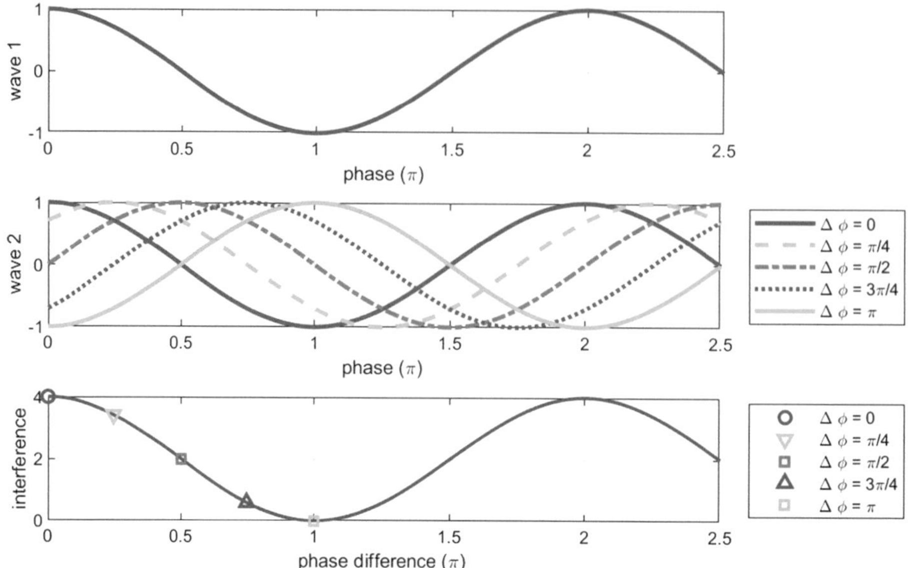

**Fig. 4.2** **a** Wave 1 versus phase, **b** Wave 2 versus phase, **c** Total intensity as a function of relative phase difference

$\Delta\theta$ is an even integer time of $\pi$ and at a minimum when $\Delta\theta$ is an odd integer time of $\pi$. The interference resulting in a maximized total intensity in this fashion is referred to as constructive interference and that results in a minimum total intensity is referred to as destructive interference.

### 4.2.2  Adding Two Waves of Different Frequencies

We now consider the situation where component waves have mutually different frequencies. In particular, we discuss the case when the frequency difference is small in detail because it leads to interesting and important concepts known as the group velocity and dispersion of waves.

**Beats**

Consider the following two waves $g_1$ and $g_2$ whose amplitude is the same and frequency is different from each other.

$$g_1(t, x) = B \sin(\omega_1 t - k_1 x) \tag{4.14}$$

$$g_2(t, x) = B \sin(\omega_2 t - k_2 x) \tag{4.15}$$

From the mathematical identity,

$$g_1 + g_2 = 2A \cos \left( \frac{(\omega_2 - \omega_1)t - (k_2 - k_1)x}{2} \right) \sin \left( \frac{(\omega_1 + \omega_2)t - (k_1 + k_2)x}{2} \right)$$

$$= 2A \cos \left( \frac{\Delta\omega}{2}t - \frac{\Delta k}{2}z \right) \sin (\omega t - kx) \tag{4.16}$$

Here, $\Delta\omega = \omega_2 - \omega_1$ and $\Delta k = k_2 - k_1$. Since $\omega_1 \simeq \omega_2$ and $k_1 \simeq k_2$, we can put $(\omega_1 + \omega_2)/2 \simeq \omega_1 = \omega$, and $(k_1 + k_2)/2 \simeq k_1 = k$. When $\omega$ and $k$ are considerably greater than $\Delta\omega$ and $\Delta k$, respectively, the cosine term on the right-hand side of expression (4.16) varies more slowly than the sine term in time and space. In this situation, we say that the two waves produce a beat and call the frequency of the slowly varying term the beat frequency.

Figure 4.3 plots an example of a beat resulting from the superposition of two component waves traveling in the positive $x$-direction. The frequencies of the first wave and the second wave are 1 Hz and 1.1 Hz, respectively whereas the phase velocity for the two waves is commonly 0.2 m/s. Since the phase velocity does not depend on the frequency we know that the two waves are assumed to be propagating through a non-dispersive medium. Figure 4.3a shows the first component wave as a function of $x$ at $t = 0$ (upper graph) and $t = 0.75$ s (lower graph). Similarly, Fig. 4.3b shows the superposed wave as a function of $x$ at $t = 0$ and $t = 0.75$ s. The waveforms in Fig. 4.3a represent the sine term of expression (4.16) at the respective times (because $\omega_1 \simeq \omega_2 = \omega$). The envelopes of the waveforms in Fig. 4.3b represent the cosine terms of expression (4.16).

It is seen that the cosine term varies much more slowly than the sine terms. From expression (4.16) it is easily known that the temporal variation of the cosine term is slower than that of the sine term as well. Notice that although the spatiotemporal variation of the two terms is different, the velocity of the component wave and that of the superposed wave is equal to each other, representing the non-dispersive propagation of the waves. The data tips in Fig. 4.3 indicate this explicitly; the labeled crest moves from 1.75 m to 1. 9 m (by 0.15 m) in 0.25 s for the component wave and 0.711 m to 0.861 m (by 0.15 m) for the superposed wave.

Expression (4.16) can be viewed as an amplitude-modulated sine wave where the sine wave propagates at the phase velocity $\omega/k$ and the amplitude modulation propagates at velocity $\Delta\omega/\Delta k$. The latter velocity represents the movement of the envelope of the wave called the wave packet.[2] The wave velocity in the form of $d\omega/dk$ (the limit case of $\Delta\omega/\Delta k$) is referred to as the group velocity and distinguished from the phase velocity $\omega/k$. In the case of (4.16), the sine term represents a wave propagating with a phase velocity. The phase velocity can be viewed as a special case of the group velocity where the temporal angular frequency $\omega$ is proportional to the spatial angular frequency $k$. When $\omega$ has second-order or higher dependence on $k$, the group velocity becomes meaningful. When the relationship between $\omega$ and $k$ is linear, the phase velocity $\omega/k$ is a constant. From this viewpoint, we can define the group velocity as the wave velocity dependent on the frequency, either the

---

[2] In Fig. 4.3 the modulated amplitude looks like a wave packet. In more general, when waves within a certain frequency band form a packet of wave.

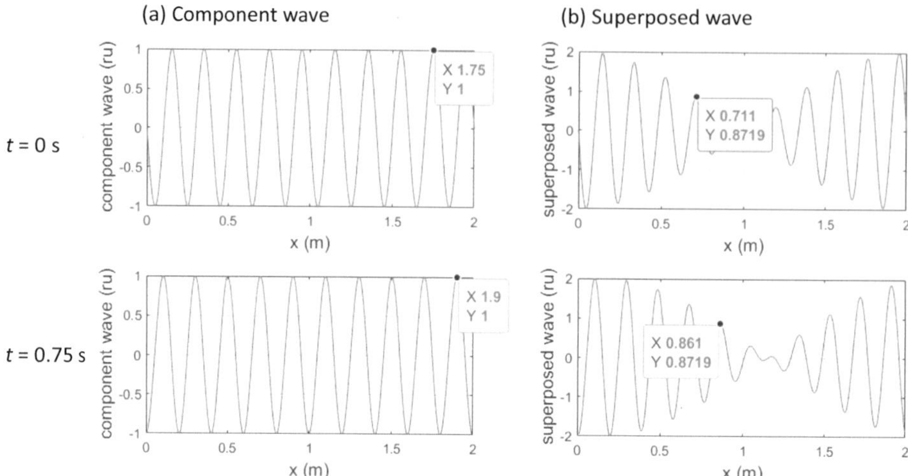



**Fig. 4.3** **a** Spatiotemporal variation of component wave and **b** superposed wave at $t = 0$ (upper) and $t = 0.75$ s (lower.) The data tip represents the location of the reference crest for each graph.

temporal frequency $\omega$ or spatial frequency $k$. The phenomenon in which the phase velocity depends on the frequency is called the dispersion. In nature, dispersion is observed when a wave propagates through a dispersive medium. Depending on the property of the dispersive medium, the group velocity can be faster or slower than the phase velocity. More on dispersion and group velocity will be discussed in Sect. 4.5.

## 4.3    Huygens' Principle

Huygens' principle states "every point on a wavefront is a source of wave" [1]. It follows that any wavefront of a given wave can be replaced by a number of small sources.

### 4.3.1    Kirchhoff's Integral Theorem

Kirchhoff formulated Huygens' principle in association with a wave equation. This formalism is known as Kirchhoff's integral theorem [5], which in the present context, can be conveniently expressed as follows.

$$\xi(x_p, y_p, z_p, t) = u(x_p, y_p, z_p)e^{-i\omega t}$$

$$= \frac{1}{4\pi} \int_S \left\{ \xi \frac{\partial}{\partial n}\left(\frac{1}{r}\right) - \frac{1}{rv}\left(\frac{\partial r}{\partial n}\right)\left(\frac{\partial \xi}{\partial t}\right) - \frac{1}{r}\left(\frac{\partial \xi}{\partial n}\right) \right\}_{t-(r/v)} dS \qquad (4.17)$$

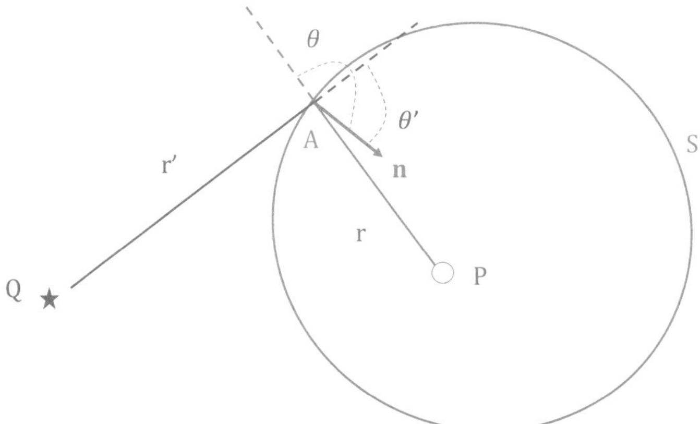

**Fig. 4.4**  Wave from point $Q$ passing through point $A$ on surface $S$ and observed at point $P$

Here, $\xi(x_p, y_p, z_p, t)$ is a wave function at a coordinate point $P(x_p, y_p, z_p)$ at time $t$. Notice that wave equations we have discussed so far such as (3.19) are in a differential form and that (4.17) takes an integral form. For derivation of expression (4.17), see Appendix A.

Consider a spherical wave in the following form as the source placed at point Q in Fig. 4.4.

$$\xi = \frac{A}{r'} e^{-i(\omega t - kr')} \tag{4.18}$$

At observation point P, the observer captures all waves passing through the surface S of the volume that encloses point P. Below we analyze the wave function at observation point P using Kirchhoff's integral theorem.

In Fig. 4.4, we consider the wave passes through at point A on the surface S. Let $r'$ be the distance from Q to A, $r$ be the distance from P to A, $\mathbf{n}$ be the inward normal vector at point $A$, $\theta$ be the angle made by vectors $\mathbf{n}$ and $\overrightarrow{PA}$, and $\theta'$ be the angle made by vectors $\mathbf{n}$ and $\overrightarrow{QA}$. Substitute (4.18) into the right-hand side of (4.17).

$$
\begin{aligned}
\xi(P, t) &= \frac{1}{4\pi} \iint_S \left\{ \frac{A}{r'} e^{-i(\omega t - kr' - \omega(r/v))} \frac{\partial}{\partial n}\left(\frac{1}{r}\right) \right. \\
&\quad + \frac{i\omega}{rv} \frac{A}{r'} e^{-i(\omega t - kr' - \omega(r/v))} \left(\frac{\partial r}{\partial n}\right) - \frac{1}{r} A e^{-i\omega(t - (r/v))} \frac{\partial}{\partial n}\left(\frac{e^{ikr'}}{r'}\right) \Bigg\} dS \tag{4.19} \\
&= \frac{1}{4\pi} \iint_S \frac{A}{rr'} e^{-i(\omega t - k(r+r'))} \left\{ \left(ik - \frac{1}{r}\right) \cos\theta - \left(ik - \frac{1}{r'}\right) \cos\theta' \right\} dS
\end{aligned}
$$

Here, from (A.21), (A.22) and (A.23) we find as follows

$$\frac{\partial r}{\partial n} = \frac{\mathbf{n} \cdot \mathbf{r}}{r} = \cos \theta \tag{4.20}$$

$$\frac{\partial}{\partial n} \left( \frac{1}{r} \right) = \frac{\partial}{\partial r} \left( \frac{1}{r} \right) \frac{\partial r}{\partial n} = -\frac{1}{r^2} \cos \theta \tag{4.21}$$

$$\frac{\partial}{\partial n} \left( \frac{e^{ikr}}{r} \right) = \frac{\partial}{\partial r} \left( \frac{e^{ikr}}{r} \right) \frac{\partial r}{\partial n} = \frac{e^{ikr}}{r} \left( ik - \frac{1}{r} \right) \cos \theta \tag{4.22}$$

Here (4.20)–(4.22) are used in going through the second equal sign in (4.19). Also used in going through the second equal sign is $v = \omega/k$.

Normally, the wavelength is much shorter than the distance from the source or observation point to point $A$, i.e., $k = 2\pi/\lambda >> r, r'$. Thus, (4.19) can be approximated as follows.

$$\xi(P, t) = \frac{ik}{4\pi} \iint_S \frac{A}{rr'} e^{-i(\omega t - k(r+r'))} \left( \cos \theta - \cos \theta' \right) dS. \tag{4.23}$$

### 4.3.2  Diffraction Due to Opening in a Plane

One of the most important applications of Huygens' principle is the description of diffraction, which is the topic of the next section. Here, using the spatial part of (4.23), $u(P)$, we discuss the basic formalism to describe diffraction due to an opening of a plate placed between a spherical wave source and an observation point. The expression obtained here is used in the next section to discuss Fraunhofer and Fresnel diffractions.

Refer to Fig. 4.5 and consider the wave from source Q passes through the opening. The distance between Q and the opening is $R'$. The observation point P is at distance $R$ on the

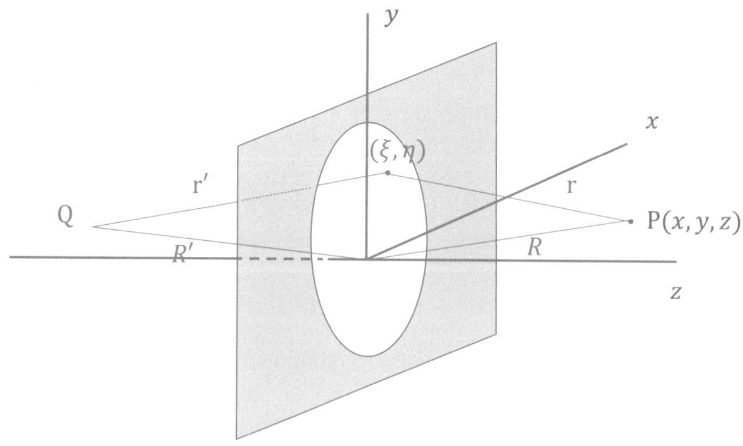


**Fig. 4.5** Wave from point $Q$ passing through an opening

other side of the opening.

$$u(P) = \frac{ik}{4\pi} \iint_S \frac{A}{rr'} e^{ik(r+r')} \left(\cos\theta - \cos\theta'\right) dS \tag{4.24}$$

When $R'$ and $R$ are much greater than the opening, the variations of $R'$, $R$, $\cos\theta$, and $\cos\theta'$ are negligibly small as compared with the variation of the phase change $exp(ik(r+r'))$ in the above surface integration. Thus, these terms can be pulled out from the integration and we can express $u(P)$ as follows.

$$u(P) \simeq \frac{ik}{4\pi} \frac{A}{RR'} \left(\cos\theta - \cos\theta'\right) \iint_S e^{ik(r+r')} dS \tag{4.25}$$

Equation (4.25) represents Huygens principle.

Now we are in a position to perform the integration. Set coordinates $(\xi, \eta)$ on the opening. We find the following relations in the coordinates.

$$\begin{aligned}
r^2 &= (x - \xi)^2 + (y - \eta)^2 + z^2 \\
r'^2 &= (x' - \xi)^2 + (y' - \eta)^2 + z'^2 \\
R^2 &= x^2 + y^2 + z^2 \\
R'^2 &= x'^2 + y'^2 + z'^2
\end{aligned} \tag{4.26}$$

So,

$$\begin{aligned}
r^2 &= R^2 - 2(x\xi + y\eta) + \xi^2 + \eta^2 \\
r'^2 &= R^2 - 2(x'\xi + y'\eta) + \xi^2 + \eta^2
\end{aligned} \tag{4.27}$$

Expanding $r$ and $r'$ to a series of powers of $\xi/R$,

$$\begin{aligned}
r &= R - \frac{x\xi + y\eta}{R} + \frac{\xi^2 + \eta^2}{2R} - \frac{(x\xi + y\eta)^2}{2R^3} - \cdots \\
r' &= R' - \frac{x'\xi + y'\eta}{R'} + \frac{\xi^2 + \eta^2}{2R'} - \frac{(x'\xi + y'\eta)^2}{2R'^3} - \cdots
\end{aligned} \tag{4.28}$$

Therefore,

$$\iint_S e^{ik(r+r')} dS = e^{ik(R+R')} \iint_S e^{ikf(\xi,\eta)} d\xi d\eta \tag{4.29}$$

where

$$\begin{aligned}
f(\xi, \eta) = &-\frac{x\xi + y\eta}{R} - \frac{x'\xi + y'\eta}{R'} + \frac{1}{2}\left(\frac{1}{R} + \frac{1}{R'}\right)(\xi^2 + \eta^2) \\
&-\frac{1}{2}\left\{\frac{(x\xi + y\eta)^2}{R^3} + \frac{(x'\xi + y'\eta)^2}{R'^3}\right\} - \cdots
\end{aligned} \tag{4.30}$$

When $R$ and $R'$ are significantly greater than $(\xi^2 + \eta^2)/\lambda$, only the first order terms in (4.30) survive. This is the case of Fraunhofer diffraction. When we need to consider up to the second-order term is the case of Fresnel diffraction.

## 4.4  Diffraction

Unlike particles, waves go around the edge of obstacles. That is why we can hear sound and see light behind an obstacle. This "going-around-corner" effect is known as the diffraction of waves. Diffraction can be understood as a combined effect of Huygens' principle and interference. When part of a wave is blocked by an obstacle, the passing part of the wavefront can be broken into small sources, and at a distance away waves from these small sources interfere with one another. Consequently, the overall wave forms a certain interference pattern. It is easily imagined that the interference pattern depends on the way the wave is blocked by the obstacle, the amplitude profile of the wavefront, and other properties of the wave. Below we discuss diffraction in general.

### 4.4.1  Fraunhofer Diffraction

When the observation point is so far that $\xi$ and $\eta$ are much smaller than $R$ and $R'$ the second and higher order terms of $\xi$ and $\eta$ can be neglected in (4.30).

$$f(\xi, \eta) \simeq -\frac{x\xi + y\eta}{R} - \frac{x'\xi + y'\eta}{R'} \tag{4.31}$$

The integral (4.29) becomes

$$\iint_S e^{ik(r+r')}dS = e^{ik(R+R')} \iint_S e^{-ik\left(\frac{x\xi + y\eta}{R} + \frac{x'\xi + y'\eta}{R'}\right)} d\xi\, d\eta \tag{4.32}$$

Accordingly, (4.25) becomes as follows. This is the case of Fraunhofer diffraction.

$$u(P) \simeq \frac{ik}{4\pi} \frac{A}{RR'} \left(\cos\theta - \cos\theta'\right) e^{ik(R+R')}$$

$$\times \iint_S e^{-ik\left(\frac{x\xi + y\eta}{R} + \frac{x'\xi + y'\eta}{R'}\right)} d\xi\, d\eta \tag{4.33}$$

Further, when $R' >> 1$ we can pull out the factor $e^{ikr'}$ from the integration in (4.25) as the variation of $r'$ on the screen (opening) is negligible. This condition corresponds to the case where the wavefront of the incident wave to the opening is flat. Thus, (4.25) can be expressed as follows.

$$u(P) \simeq \frac{ik}{4\pi} \frac{A}{RR'} \left(\cos\theta - \cos\theta'\right) e^{ikR'} \iint_S e^{-ikr} dS$$

$$= \frac{ik}{4\pi} \frac{A}{RR'} \left(\cos\theta - \cos\theta'\right) e^{ik(R'+R)} \iint_S e^{-ik(l\xi+m\eta)} d\xi d\eta \qquad (4.34)$$

Here $x/R = l$, $y/R = m$.

## 4.4.2  Fresnel Diffraction

In this case, we cannot neglect the second-order terms of $\xi$ and $\eta$.

$$u(P) \simeq \frac{ik}{4\pi} \frac{A}{RR'} e^{ik(R+R')} \left(\cos\theta - \cos\theta'\right) \iint_S e^{ikf(\xi,\eta)} d\xi d\eta \qquad (4.35)$$

$$f(\xi,\eta) = -\frac{x\xi + y\eta}{R} - \frac{x'\xi + y'\eta}{R'} + \frac{1}{2}\left(\frac{1}{R} + \frac{1}{R'}\right)(\xi^2 + \eta^2)$$

$$-\frac{1}{2}\left\{\frac{(x\xi + y\eta)^2}{R^3} + \frac{(x'\xi + y'\eta)^2}{R'^3}\right\} \qquad (4.36)$$

The complexity is the evaluation of the surface integral term in (4.35). Using Euler's notation, put the integral in the following form.

$$\iint_S e^{ikf(\xi,\eta)} d\xi d\eta = C + iS \qquad (4.37)$$

where

$$C = \iint_S \cos\{kf(\xi,\eta)\} d\xi d\eta \qquad (4.38)$$

$$S = \iint_S \sin\{kf(\xi,\eta)\} d\xi d\eta \qquad (4.39)$$

Using the following variable conversions (Fig. 4.6),

**Fig. 4.6** Fresnel diffraction

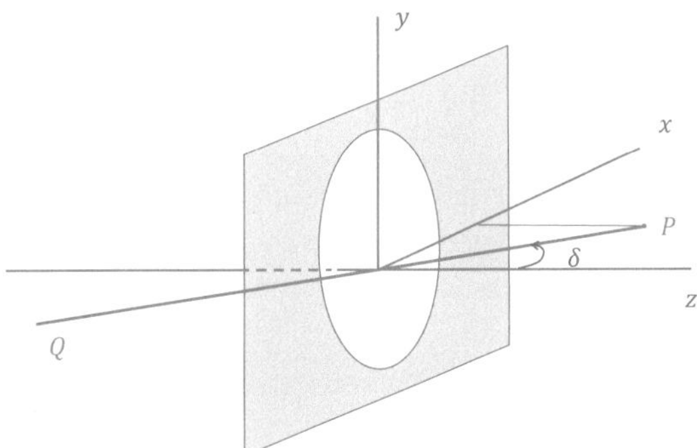

$$\frac{x}{R} = \sin\delta, \qquad \frac{x'}{R'} = -\sin\delta, \tag{4.40}$$

$$\frac{y}{R} = 0, \qquad \frac{y'}{R'} = 0, \tag{4.41}$$

$$\frac{z}{R} = \cos\delta, \qquad \frac{z'}{R'} = -\cos\delta, \tag{4.42}$$

By substituting (4.40) and (4.41) into (4.36) we can simplify the expression of $f(\xi, \eta)$ as follows.

$$
\begin{aligned}
f(\xi, \eta) &= \frac{1}{2}\left(\frac{1}{R} + \frac{1}{R'}\right)(\xi^2 + \eta^2) - \frac{1}{2}\left(\frac{1}{R} + \frac{1}{R'}\right)\xi^2 \sin^2\delta \\
&= \frac{1}{2}\left(\frac{1}{R} + \frac{1}{R'}\right)(\xi^2 \cos^2\delta + \eta^2)
\end{aligned}
\tag{4.43}
$$

Using this expression of $f(\xi, \eta)$ in expression of $C$ (4.38) and $S$ (4.39) and putting the term multiplied to the integration in $u(P)$ (4.35) as follows,

$$B = \frac{ik}{4\pi}\frac{A}{RR'}e^{ik(R+R')}\left(\cos\theta - \cos\theta'\right)$$

we obtain the following expression.

$$I(P) = |B|^2\left(C^2 + S^2\right) \tag{4.44}$$

Expression (4.44) represents the intensity of the diffraction pattern at point $P$.


### 4.4.3  Diffraction Due to Rectangular Opening

Consider Fraunhofer diffraction due to a rectangular opening as an example. For simplicity, use (4.34) (a wave with a flat wavefront incident to the opening). Since our interest is the diffraction pattern on the screen, consider the spatial integration part of this equation.

$$
\begin{aligned}
\iint_S e^{ikr}\,dS &= e^{ikR}\int_{-a}^{a}\int_{-b}^{b} e^{-ik(l\xi+m\eta)}\,d\xi\,d\eta \\
&= e^{ikR}\int_{-a}^{a} e^{-ikl\xi}\,d\xi \int_{-b}^{b} e^{-ikm\eta}\,d\eta = e^{ikR}\frac{2a\sin kla}{kla}\frac{2b\sin kmb}{kmb}
\end{aligned}
\tag{4.45}
$$

The opening is in the area $-a \le x \le a$ and $-b \le y \le b$. Put $kal = \alpha$ and $kmb = \beta$ to express the integral as follows.

$$\iint_S e^{-ik(l\xi+m\eta)}\,d\xi\,d\eta = \left(\frac{2a\sin\alpha}{\alpha}\right)\left(\frac{2b\sin\beta}{\beta}\right) \tag{4.46}$$

From (4.34) and (4.46), the amplitude of the diffraction pattern at point $P(x, y)$ is written in the following form.

$$u(P) = B_0 \left( \frac{2a \sin \alpha}{\alpha} \right) \left( \frac{2b \sin \beta}{\beta} \right) \tag{4.47}$$

Here $B_0$ is

$$B_0 = \frac{ik}{4\pi} \frac{A}{RR'} e^{ik(R'+R)} \left( \cos \theta - \cos \theta' \right) \tag{4.48}$$

Further, the intensity $I(x, y)$ can be found as the square of expression (4.47).

$$I(x, y) = B_0^2 \left( \frac{2a \sin \alpha}{\alpha} \right)^2 \left( \frac{2b \sin \beta}{\beta} \right)^2 \tag{4.49}$$

## 4.5   Dispersion

The phenomenon that the wave (phase) velocity depends on the wave number $k$ is referred to as dispersion. Since the wave number is related to the frequency $v$ as $v_p = \omega/k = v\lambda$ ($\lambda$ is the wavelength), we may say that dispersion is the phenomenon that the wave velocity depends on the frequency. When a wave exhibits dispersion as it passes through a medium, we say that the medium is dispersive.

Dispersion becomes important when a wave packet, especially a pulsed wave, travels through a dispersive medium. As we discuss in Appendix (B), a pulse contains a band of frequency. If a pulse travels through a dispersive medium at a certain group velocity, each frequency component has a different phase velocity. Therefore, the pulse width increases as the pulsed wave travels. When a train of pulses travels with a certain interval between the neighboring pulses, this can cause the leading edge of a pulse to catch up with the trailing edge of the pulse ahead of it. Consequently, the information carried by the two pulses can interfere with each other. Generally, this is an unfavorable phenomenon. Thus, managing the dispersive characteristics of a medium is important.

### 4.5.1   Group Velocity

Dispersion is observed when the material's property is such that the phase velocity depends on the frequency. Normally, the frequency of a wave is determined by the source and the phase velocity is determined by the material through which the wave travels. In a realistic situation, the frequency of the incidence wave has some width (called the bandwidth) as any realistic source has some mechanism of spectral broadening. When a wave of frequency with a certain bandwidth travels through a dispersive medium, all frequency components travel at different phase velocities. This causes two practical effects. First, the shape of the wave changes as the wave travels (that is why the phenomenon is referred to as dispersion). Second, the phase velocity becomes less meaningful as compared with the case when the wave does not show dispersion. In this case, group velocity is used. We will discuss this concept below.

The group velocity $v_g$ can be related to the phase velocity $v_p$ as follows.

$$v_p(k) = \frac{\omega}{k} \tag{4.50}$$

$$\omega(k) = v_p(k)k \tag{4.51}$$

$$v_g(k) = \frac{d\omega}{dk} = v_p + \frac{dv_p}{dk}k \tag{4.52}$$

Here, $\omega$ and $k$ are the angular frequency and the wave number. If the phase velocity is a decreasing function of $k$, the $dv_p/dk < 0$ term on the right-hand side of (4.52) makes $v_g < v_p$. This type of dispersion is referred to as normal dispersion. If the phase velocity is an increasing function of $k$, $v_g > v_p$. In this case, the dispersion is said to be anomalous. In the case of light, $v_p = c/n(k)$. Here, $n$ is the index of refraction as a function of wave number. Normal dispersion occurs in the frequency range where the index of refraction is an increasing function of $\omega$, and anomalous dispersion occurs in the frequency range where the index of refraction decreases with $\omega$.

Conventionally, normal and anomalous dispersion are discussed with a wave number frequency diagram. Figure 4.7 illustrates sample normal and anomalous dispersion. Consider the phase velocity at a reference wave number $k_r$.

$$v_p = \left(\frac{d\omega}{dk}\right)_{k=k_r} \tag{4.53}$$

Using (4.53), we can put as follows.

$$\frac{dv_p}{dk} = \left(\frac{d^2\omega}{dk^2}\right)_{k=k_r} \tag{4.54}$$

In Fig. 4.7a the $\omega$-$k$ curve is convex upward or $d^2\omega/dk^2 = dv_p/dk < 0$, and according to (4.52) $v_g < v_p$. In (b), $d^2\omega/dk^2 > 0$ and therefore $v_g > v_p$. In terms of the frequency dependence of the phase velocity, normal dispersion is the case where the phase velocity decreases with the frequency, and anomalous dispersion is the case where the phase velocity increases with the frequency. Figure 4.7 illustrates the situation at angular frequencies $\omega_1$ and $\omega_2 > \omega_1$.

Figures 4.8, 4.9 and 4.10 illustrate the relation between phase velocity and group velocity, using a simple wave consisting of two component waves (wave 1 and wave 2). Both component waves are sinusoidal waves where wave 1's frequency is 0.9 Hz and wave 2's frequency is 1.0 Hz. It is considered that the superposed wave travels through three types of media under different dispersive conditions. Case 1 is when the medium is not dispersive. Case 2 and Case 3 are when the medium has normal dispersion and anomalous dispersion, respectively, in the present frequency range. Table 4.1 summarizes these conditions. Since the two component waves have frequencies close to each other the superposed wave exhibits a pattern of beat.

**Table 4.1** Conditions of two waves being superposed; $v_{p1}$ and $v_{p2}$ are phase velocity at 0.9 Hz and 1.0 Hz.

|  | $v_{p1}$ (m/s) | $v_{p2}$ (m/s) | $v_g$ (m/s) | $dv_p/d\omega$ | dispersion |
|---|---|---|---|---|---|
| Case 1 | 0.200 | 0.200 | 0.200 | 0 | No dispersion |
| Case 2 | 0.222 | 0.200 | 0.106 | < 0 | Normal dispersion |
| Case 3 | 0.178 | 0.200 | -1.78 | > 0 | Anomalous dispersion |

The group velocity can be found in the movement of the beat pattern, as discussed below.

**Case 1: No Dispersion**

In this case, the medium is non-dispersive and the phase velocity is 0.2 m/s both at 0.9 Hz and 1.0 Hz. Figure 4.8 shows the superposed wave at $t = 0, 5$ and 10 s. Due to the frequency difference, the superposed wave shows the pattern of the beat. Since the medium is not dispersive, the group velocity should be equal to the phase velocity. We can use the beat pattern to estimate the group velocity and verify that it is equal to the phase velocity. The beat pattern moves in 5 s by 1 m as the dashed squares in Fig. 4.8 indicate, i.e., the center of the dashed square moves from 2 m to 3 m in the first 5 s, and 3 m to 4 m in the second 5 s; $1\,\text{m} \div 5\,\text{s} = 0.2\,\text{m/s}$. Indeed it is the same as the phase velocity 0.2 m/s.

**Case 2: Normal Dispersion**

In this case, the medium is dispersive such that the phase velocity is 0.222 m/s at 0.9 Hz and 0.200 m/s at 1.0 Hz. Since the phase velocity decreases with frequency the dispersion is normal. Figure 4.9 indicates that the beat pattern enclosed by the dashed square moves approximately 0.5 m in 5 s, hence the group velocity is estimated to be 0.1 m/s. This agrees with the group velocity calculated based on $d\omega/dk$ (4.52) and shown in Table 4.1.

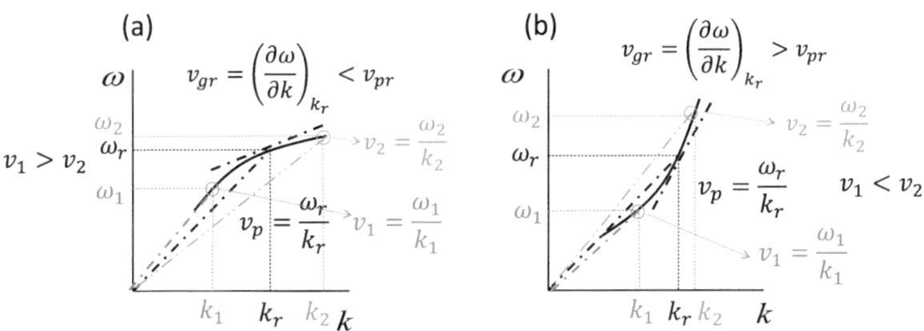


**Fig. 4.7** **a** Normal dispersion and **b** anomalous dispersion

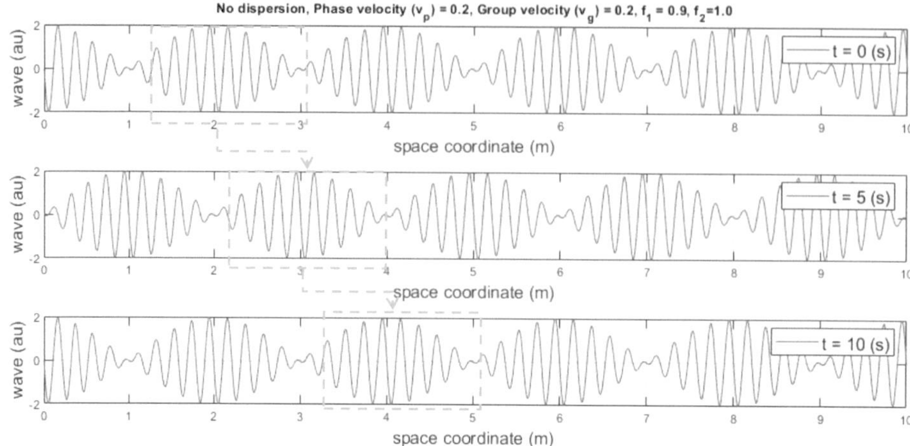



**Fig. 4.8** Two sinusoidal waves having similar frequencies traveling through a medium with no dispersion

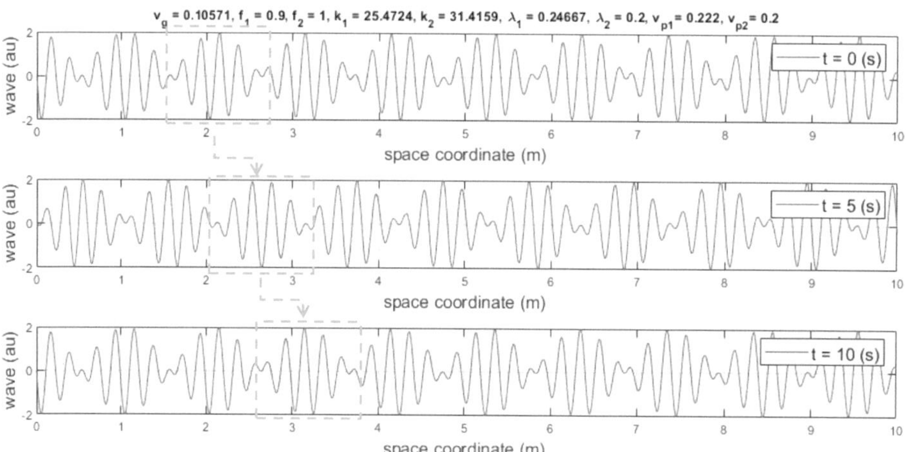



**Fig. 4.9** Two sinusoidal waves having similar frequencies traveling through a medium with normal dispersion

## Case 3: Anomalous Dispersion

The phase velocity, in this case, is 0.178 m/s at 0.9 Hz and 0.200 m/s at 1.0 Hz, hence the dispersion is anomalous. Figure 4.10 indicates that the beat pattern moves backward by approximately 1.8 m in 1 s, i.e., the group velocity is -1.8 m/s. This agrees with the group velocity calculated based on $d\omega/dk$ (4.52) and shown in Table 4.1. The group velocity appears to be negative because the beat pattern moves much faster than the phase velocity of 0.200 m/s. Figure 4.11 shows the superposed wave at the same time steps as Fig. 4.10 in

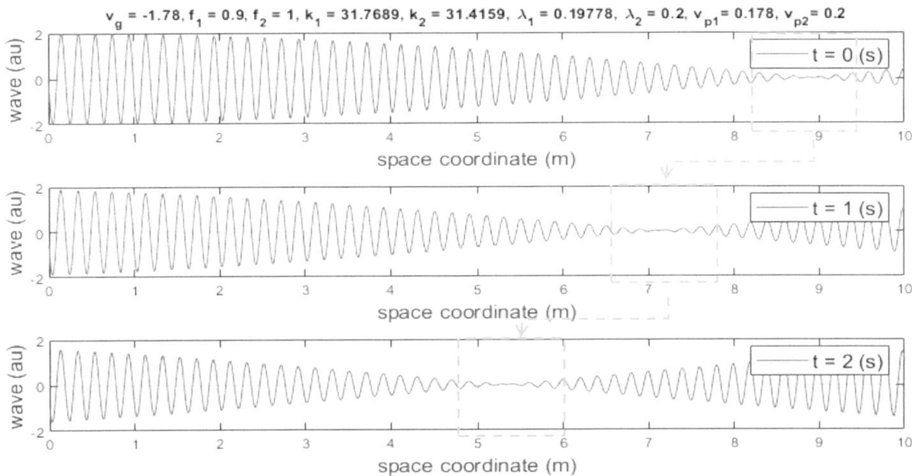



**Fig. 4.10** Two sinusoidal waves having similar frequencies traveling through a medium with anomalous dispersion

a larger time window. It is explicitly seen that the beat pattern moves approximately 14 m in 1 s, indicating that the group velocity is much faster than the phase velocity of 0.200 m/s.

As discussed above, we can interpret dispersion as the phenomenon that the phase velocity depends on the frequency due to the dispersive property of the material that the wave propagates through. Figure 4.12 is a phase velocity as a function of frequency illustrating the three dispersive situations discussed in Figs. 4.8, 4.9 and 4.10.

### 4.5.2 Sample Dispersive Systems

In this section, we consider two examples of dispersive physical systems. The first one is an elastic medium with a damping mechanism and the second is a simple spring-mass system connected to an elastic medium.

**Elastic Medium with Damping Mechanism**

In Sect. 3.2 we discussed the one-dimensional wave equation (3.26) for the displacement field in an elastic medium. We derived a wave solution (3.38) expressed with a real temporal frequency $\omega$ and complex spatial frequency $k_r + i k_i$, and discussed that the phase velocity evaluated as the ratio of the temporal frequency to the real part of the spatial frequency $v_p = \omega/k_r$ is a function of $\omega$. The phase velocity expression (3.39) indicates that this $\omega$ dependence is in the form of $2\beta/\omega$.

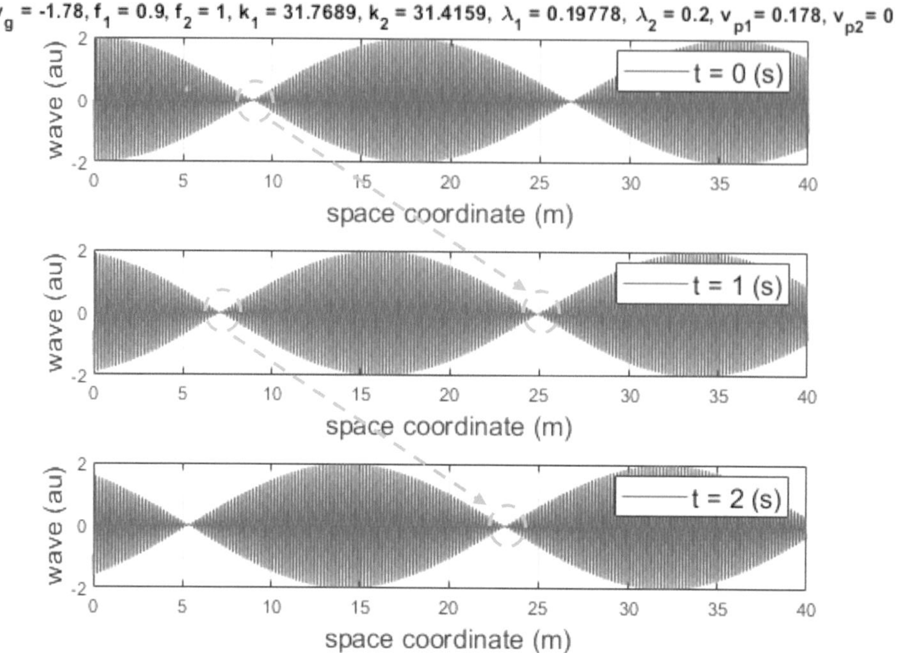



**Fig. 4.11** Two sinusoidal waves having similar frequencies traveling through a medium with anomalous dispersion for a wider range. Close-up-view

$$v_p = \frac{\omega}{k_r} = \sqrt{\frac{2E}{\rho}}\left[1 + \sqrt{1 + \left(\frac{2\beta}{\omega}\right)^2}\right]^{-\frac{1}{2}} \qquad (3.39)$$

Expression (3.39) indicates that if the decay constant $\beta$ is zero the phase velocity becomes a constant value. This observation indicates that the dispersion is caused by the decay (damping) characteristics of the elastic system.

It is interesting to note that the solution to the same wave equation (3.26) can be put in the form where the temporal frequency is complex.

$$\xi_x(t, x) = \xi_0 e^{-\omega_i x} e^{i(\omega_r t \pm kx)} \qquad (4.55)$$

where the real and imaginary parts of the complex temporal frequency are

$$\omega_r = \sqrt{\frac{E}{\rho}k^2 - \beta^2} \qquad (4.56)$$

$$\omega_i = \beta \qquad (4.57)$$

In this case, the phase velocity becomes as follows.

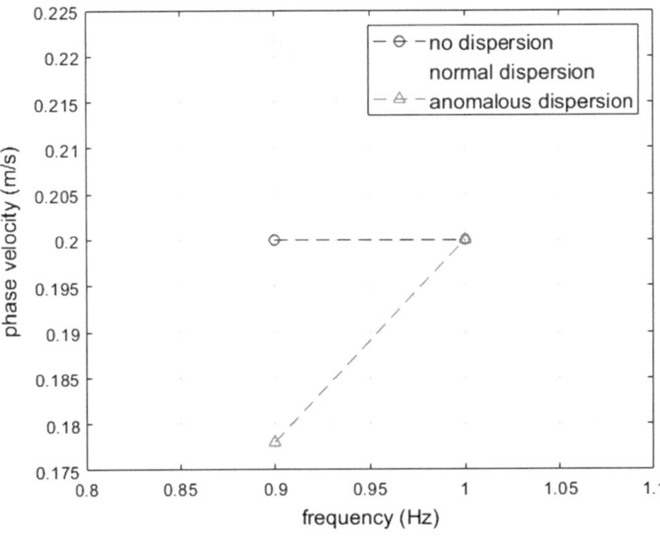

**Fig. 4.12** Frequency dependence of phase velocity for the three sample cases of dispersion discussed in Figs. 4.8, 4.9 and 4.10

$$v_p = \frac{\omega}{k} = \sqrt{\frac{E}{\rho} - \frac{\beta}{k}} \tag{4.58}$$

The phase velocity is an explicit function of the spatial frequency $k$. As expected, if $\beta = 0$ the phase velocity becomes independent of $k$.

**String on Elastic Medium**

The second example of a dispersive system is a combination of an elastic string and an elastic substrate to which the string is attached. Figure 4.13 schematically illustrates the system.

In Sect. 3.1.1 we considered the transverse wave propagating on a string and derived wave equation (3.7).

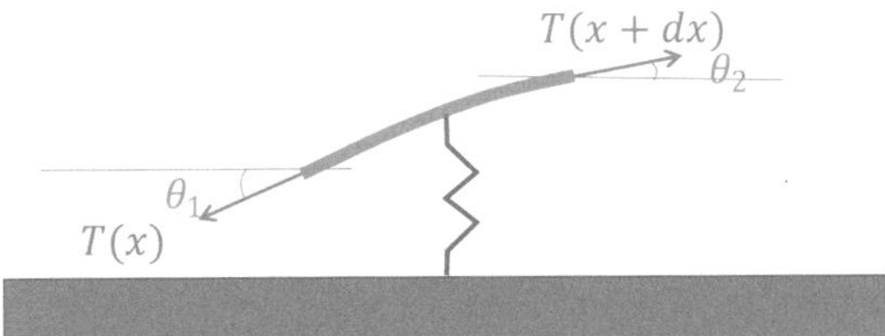


**Fig. 4.13** String attached to an elastic medium. The spring represents the elastic medium the string is attached to. The elastic medium can be part of the substrate the string is attached to or an intermediate elastic medium sandwiched by the string and the substrate

$$\frac{d^2\xi_y}{dt^2} = \frac{T}{\mu}\frac{d^2\xi_y}{dx^2} \qquad (3.7)$$

Now consider that the same string is glued to the surface of an elastic medium of stiffness $K$. The elastic medium exerts vertical elastic force $(K/l)\xi_y$ on the unit length of the string. This action adds the vertical force term to (3.7) as follows.

$$\frac{d^2\xi_y}{dt^2} = \frac{T}{\mu}\frac{d^2\xi_y}{dx^2} - \frac{K/l}{\mu}\xi_y = \frac{T}{\mu}\frac{d^2\xi_y}{dx^2} - \frac{K}{m}\xi_y \qquad (4.59)$$

Here $m$ and $l$ are the mass and length of the string. The characteristic equation of differential equation (4.59) yields the following relation between the temporal frequency $\omega$ and spatial frequency $k$.

$$\omega = \sqrt{\frac{T}{\mu}k^2 + \frac{K}{m}} \qquad (4.60)$$

From (4.60), we find the following expression of phase velocity.

$$v_p(k) = \frac{\omega}{k} = \sqrt{\frac{T}{\mu} + \frac{K}{mk^2}} \qquad (4.61)$$

Using (4.60), we can express the phase velocity as an explicit function of $\omega$ as well.

$$v_p(\omega) = \frac{\omega}{k} = \sqrt{\frac{T}{\mu}\left(\frac{m\omega^2}{m\omega^2 + K}\right)} \qquad (4.62)$$

Since the phase velocity depends on the frequency $k$ or $\omega$, this system is dispersive. It is a one-dimensional version of a thin-film system such as a metal-film coated on a silicon substrate. In this system, the metal-silicon interface behaves as an elastic medium. From the dispersion characteristics, it is possible to evaluate the adhesion characteristics [6].

## References


1. Hecht E (2002) Optics, 4th edn. Addison Wesley, San Francisco, CA, USA. Chap. 4
2. Yoshida S (2023) Fundamentals of optical waves and lasers. Springer Nature, Cham, Switzerland
3. Kinsler LE, Frey AR, Coppens AB, Sanders JV (1980) Fundamentals of acoustics, 3rd edn, , Wiley, New York, USA. Chap. 6
4. Graff KF (1975) Wave motion in elastic solids. Oxford University Press, Oxford, UK
5. Kirchhoff G (1883) Zur Theorie der Lichtstrahlen. Ann d Physik 2(18):663–695
6. Park IK, Park TS, Miyasaka C (2012) Quantitative NDE of the nano-scaled thin film system using SAM. In: 2012 IEEE international ultrasonics symposium, Dresden, pp 1742–1745.https://doi.org/10.1109/ULTSYM.2012.0437

# Fourier Series and Fourier Transform  5

Often, we better explain oscillation and wave dynamics in the frequency domain. In this and the following chapter, we discuss oscillation and wave dynamics in the frequency domain. Each physical system exhibits its unique frequency response to input. This response, known as the transfer function, characterizes the system. We can describe the behavior of a linear system consisting of multiple unit systems as the product of the transfer functions of the unit systems.

As the first step, we focus on the Fourier Transform and its properties in this chapter. We start this chapter with the formalism of the Fourier Transform. After discussing the mathematical meanings of some aspects of the Fourier Transform using examples, we discuss several properties of the Fourier Transform. The contents of this chapter are applicable to describe linear systems' frequency responses. Appendix B lists the Fourier Transforms for several, typical functions.

## 5.1 Fourier Series

According to Fourier's theorem [1–3] a periodic function can be expanded into a series of cosine and sine functions.

$$f(t) = \frac{a_0}{2} + a_1 \cos(\omega_0 t) + a_2 \cos(2\omega_0 t) + \cdots + a_n \cos(n\omega_0 t)$$
$$+ b_1 \sin(\omega_0 t) + b_2 \sin(2\omega_0 t) + \cdots + b_n \sin(n\omega_0 t) \tag{5.1}$$

Here $\omega_0$ is the lowest angular frequency. If function $f(t)$ is defined on $[0 \ T]$, the lowest frequency $\nu_0$ and the corresponding angular frequency are as follows.

© The Author(s), under exclusive license to Springer Nature Switzerland AG 2025     75
S. Yoshida, *Physics and Mathematics Behind Wave Dynamics*, Synthesis Lectures
on Wave Phenomena in the Physical Sciences,
https://doi.org/10.1007/978-3-031-60354-9_5

$$v_0 = \frac{1}{T} \tag{5.2}$$

$$\omega_0 = 2\pi v_0 = \frac{2\pi}{T} \tag{5.3}$$

With this notation, we can express (5.1) as follows.

$$
\begin{aligned}
f(t) &= \frac{a_0}{2} + \sum_{k=1}^{n} \left\{ a_k \cos\left(\frac{2\pi k}{T} t\right) + b_k \sin\left(\frac{2\pi k}{T} t\right) \right\} \\
&= \frac{a_0}{2} + \sum_{k=1}^{n} \left\{ a_k \cos(\omega_k t) + b_k \sin(\omega_k t) \right\}
\end{aligned}
\tag{5.4}
$$

Once we express the function $f(t)$ as the series of cosine and sin functions, the next step is to determine the coefficients $a_n$ and $b_n$. We can use the following property to determine the coefficients. This property is known as the orthogonality of the trigonometric functions [4, 5].

$$\int_{t_1}^{t_1+T} \cos(n\omega_0 t)\cos(m\omega_0 t)dt = 0 \quad (n \neq m) \tag{5.5}$$

$$= \frac{T}{2} = \frac{\pi}{\omega_0} \quad (n = m) \tag{5.6}$$

Here the following identities are used.

$$\cos(n\omega_0 t)\cos(m\omega_0 t) = \frac{1}{2}\{\cos((m-n)\omega_0 t) + \cos((m+n)\omega_0 t)\} \tag{5.7}$$

$$\sin(n\omega_0 t)\sin(m\omega_0 t) = \frac{1}{2}\{\cos((m-n)\omega_0 t) - \cos((m+n)\omega_0 t)\} \tag{5.8}$$

$$\cos^2(n\omega_0 t) = \frac{1}{2}\{1 + \cos(2n\omega_0 t)\} \tag{5.9}$$

$$\sin^2(n\omega_0 t) = \frac{1}{2}\{1 - \sin(2n\omega_0 t)\} \tag{5.10}$$

Multiply $\cos(2\omega_0 t)$ to both sides of (5.1) and integrate for one period (from $t = t_1$ to $t = t_1 + T$). From the orthogonality, all terms vanish except $a_2 \cos(2\omega_0 t)$ on the right-hand side. Thus, using (5.6), we find that coefficient $a_2$ can be expressed as follows.

$$a_2 = \frac{2}{T} \int_{t_1}^{t_1+T} f(t)\cos(2\omega_0 t)dt \tag{5.11}$$

It is clear that other coefficients can be evaluated with the same procedure. Generally, coefficients $a_n$ and $b_n$ can be expressed as follows.

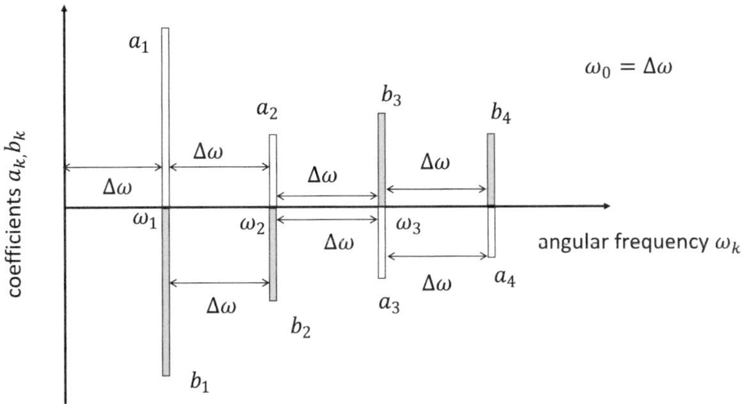


**Fig. 5.1**  Coefficients of Fourier series

$$a_n = \frac{2}{T} \int_{t_1}^{t_1+T} f(t)\cos(n\omega_0 t)\,dt \tag{5.12}$$

$$b_n = \frac{2}{T} \int_{t_1}^{t_1+T} f(t)\sin(n\omega_0 t)\,dt \tag{5.13}$$

Figure 5.1 illustrates the coefficients schematically.

Now consider increasing $T$ in the range $[0\ T]$. As $T$ increases, the frequency step $\Delta\omega$ decreases. As the frequency step decreases the coefficient series $a_k$, $b_k$ becomes a continuous function of $\omega$. It is convenient to define spectral densities $A(\omega)$ and $B(\omega)$ as follows.

$$A(\omega_k) = \frac{a_k}{\Delta\omega} \tag{5.14}$$

$$B(\omega_k) = \frac{b_k}{\Delta\omega} \tag{5.15}$$

Using (5.14) and (5.15), we can write (5.4) as follows.

$$f(t) = \frac{a_0}{2} + \sum_{k=1}^{n} \{A(\omega_k)\cos(\omega_k t) + B(\omega_k)\sin(\omega_k t)\}\,\Delta\omega \tag{5.16}$$

In the limit $T \to \infty$, the frequency step becomes infinitesimal as $\Delta\omega = \omega_{k+1} - \omega_k \to d\omega$. In this limit, the summation (5.16) becomes an integral.

$$f(t) = \frac{a_0}{2} + \int_{k=1}^{n} \{A(\omega)\cos(\omega t) + B(\omega)\sin(\omega t)\}\,d\omega \tag{5.17}$$

The right-hand side of (5.17) is called the Fourier integral. At this point, it becomes clear that the Fourier series in the form of (5.17) represents the transform between the functions of frequency $A(\omega)$, $B(\omega)$, and the function of time $f(t)$. The transform from a function of time to a function of frequency is called the Fourier transform and the opposite is called the inverse Fourier transform. We will discuss more details of the Fourier transform later in this section.

## 5.2    Exponential Form of Fourier Series

Using Euler's notation (1.19), we can compactly express the Fourier series. Applying (1.19) to the $n\omega_0 t$ terms of (5.1), we find as follows. Here $j$ is the imaginary unit ($j^2 = -1$).

$$\cos n\omega_0 t = \frac{e^{jn\omega_0 t} + e^{-jn\omega_0 t}}{2} \tag{5.18}$$

$$\sin n\omega_0 t = \frac{e^{jn\omega_0 t} - e^{-jn\omega_0 t}}{2j} \tag{5.19}$$

Use (5.18) and (5.19) in (5.1).

$$
\begin{aligned}
f(t) &= a_0 + a_1 \frac{e^{j\omega_0 t} + e^{-j\omega_0 t}}{2} + a_2 \frac{e^{j2\omega_0 t} + e^{-j2\omega_0 t}}{2} + \cdots + a_n \frac{e^{jn\omega_0 t} + e^{-jn\omega_0 t}}{2} \\
&+ b_1 \frac{e^{j\omega_0 t} - e^{-j\omega_0 t}}{2j} + b_2 \frac{e^{j2\omega_0 t} - e^{-j2\omega_0 t}}{2j} + \cdots + b_n \frac{e^{jn\omega_0 t} - e^{-jn\omega_0 t}}{2j} \\
&= a_0 + (\frac{a_1}{2} - \frac{jb_1}{2})e^{j\omega_0 t} + \cdots + (\frac{a_n}{2} - \frac{jb_n}{2})e^{jn\omega_0 t} \\
&+ (\frac{a_1}{2} + \frac{jb_1}{2})e^{-j\omega_0 t} + \cdots + (\frac{a_n}{2} + \frac{jb_n}{2})e^{-jn\omega_0 t}
\end{aligned}
\tag{5.20}
$$

Defining coefficient $c_n$ as follows,

$$c_n = \frac{1}{2}(a_n - jb_n) \tag{5.21}$$

$$c_n^* = \frac{1}{2}(a_n + jb_n) \tag{5.22}$$

we can put (5.20) in the following form.

$$
\begin{aligned}
f(t) &= c_0 + c_1 e^{j\omega_0 t} + c_1^* e^{-j\omega_0 t} + \cdots + c_n e^{jn\omega_0 t} + c_n^* e^{-jn\omega_0 t} \\
&= \sum_{n=0}^{\infty} (c_n e^{jn\omega_0 t} + c_n^* e^{-jn\omega_0 t})
\end{aligned}
\tag{5.23}
$$

Further, if we define $c_{-n}$ as

$$c_{-n} \equiv c_n^* \tag{5.24}$$

we can express (5.23) and (5.21) in a compact form as follows.

$$f(t) = \sum_{n=-\infty}^{\infty} c_n e^{jn\omega_0 t} \tag{5.25}$$

$$c_n = \frac{1}{T} \int_{t_1}^{t_1+T} f(t)e^{-jn\omega_0 t}\,dt \tag{5.26}$$

We can interpret the above expressions as follows. When the index $n$ is negative, the frequency $n\omega_0$ becomes negative on the right-hand side in the expression of the corresponding coefficient $c_n$, (5.26). According to (5.24), the coefficient $c_n$ with a negative $n$ is the complex conjugate ($c_n^*$) of its positive-frequency counterpart.

The above expressions yield symmetric patterns in the frequency-domain expressions. Figure 5.2 shows the symmetric patterns using the Fast Fourier Transform (FFT) [1, 6] of a square wave function, as an example. Here, Fig. 5.2a shows the magnitude of the FFT expressed on the frequency axis ranging from the lowest frequency defined by (5.2) to the sampling frequency, $F_s$. According to Nyquist theorem [7, 8] the physically meaningful maximum frequency is half of the sampling frequency, $F_s/2$. Thus, we can define the negative

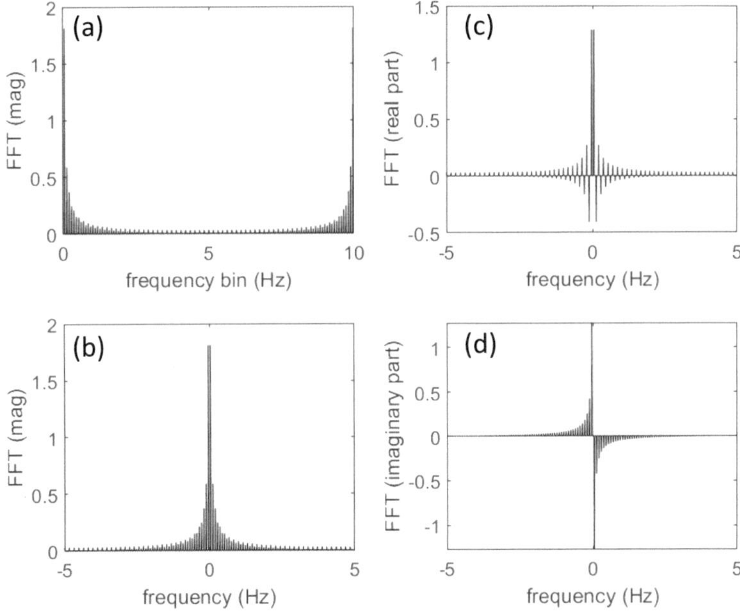

**Fig. 5.2** FFT with negative frequency. **a** Output of FFT magnitude; **b** FFT magnitude on negative and positive frequency; **c** Real part of FFT; **d** Imaginary part of FFT

frequency as ranging from $-F_s/2$ Hz to $0$ Hz and the positive frequency as ranging from $0$ Hz to $F_s/2$ Hz. Figure 5.2b rearranges the FFT magnitude of Fig. 5.2a on the frequency axis ranging from $-F_s/2$ Hz to $F_s/2$ Hz.

Figure 5.2c and d show the real and imaginary parts of the FFT. Since the FFT on the negative side of the frequency axis is the complex conjugate of the FFT on the positive side of the frequency axis, the real part is symmetric about the vertical axis and the imaginary part is symmetric about the origin of the graph (the point where the vertical axis crosses $0$ Hz). Figure 5.2c and d indicate these symmetries.

It is important to know that the complex conjugate part is necessary to compute the inverse Fourier Transform (i-FFT). Even missing one term in the complex conjugate, e.g., the omission of the term corresponding to $-f_k$ Hz with the inclusion of the complex conjugate at $+f_k$ Hz, results in a complex number in the resultant i-FFT. In other words, the time series resulting from the i-FFT becomes a complex number. A complex time series does not represent a single signal. Remember that as we already discussed, the imaginary part represents a phase shift of a quarter period. Since we started with a single square wave, if the i-FFT signal is a mixture of a signal and a quarter-period shifted signal, the operation is physically incorrect.

Consider the meaning of the minimum angular frequency $\omega_0$ in (5.25). What we try to do is to expand a certain function of time into a series of cosine and sine functions. Here the minimum frequency $\nu_0 = \omega_0/2\pi$ corresponds to the situation where the entire length of the given function is expressed with $\cos 2\pi \nu_0 t$ and $\sin 2\pi \nu_0 t$. In other words, we assume that the entire time duration in which the function is defined is equal to the longest (fundamental) period of the series ($T$). The second term $\cos 2\omega_0 t$ and $\sin 2\omega_0 t$ correspond to the period half of the fundamental period ($T/2$). In general, the $n^{\text{th}}$ frequency corresponds to (n $T/N$). Here N is determined by the total duration and the time step with which the data is taken; N = T/$\Delta$t.

## 5.3 Fourier Transform

We can extend the above concept to a general case where the function is not periodic. To this end, we modify the above argument for periodical functions by making the period $T$ infinitely long in (5.26). Below we consider the Fourier series under this limit case.

Rewriting the relation between the minimum angular frequency and the period given by (5.3) in the unit of Hz and s, we can relate the minimum frequency and the period as $\Delta \nu = 1/T$. Substituting (5.26) into (5.25), and replacing $1/T$ with $\Delta \nu$, and $n\omega_0$ with $n2\pi \nu_0$, we obtain the following expression.

$$f(t) = \sum_{n=-\infty}^{\infty} c_n e^{jn2\pi \nu_0 t} = \sum_{n=-\infty}^{\infty} \left\{ \int_{t_1}^{t_1+T} f(t) e^{-jn2\pi \nu_0 t} dt \right\} e^{jn2\pi \nu_0 t} \Delta \nu \qquad (5.27)$$

Consider the limit case where the period $T$ is infinitely long. Under this condition, the limit of the integration in the curly bracket in (5.27) becomes from $-\infty$ to $\infty$, and the frequency bin width ($\Delta v$) becomes infinitesimally small ($dv$). It follows that $n2\pi v_0 = 2\pi v$. Accordingly, we can replace the summation with an integral and rewrite (5.27) as follows.

$$f(t) = \int_{-\infty}^{\infty} \left\{ \int_{-\infty}^{\infty} f(t)e^{-j2\pi vt}\,dt \right\} e^{j2\pi vt}\,dv \tag{5.28}$$

The content of the curly bracket in (5.28) is known as the Fourier transform of the function $f(t)$. Using $\mathcal{F}$ as the symbol for Fourier transform, we can put (5.28) in the following form.

$$\mathcal{F}\{f(t)\} = F(v) = \int_{-\infty}^{\infty} f(t)e^{-j2\pi vt}\,dt \tag{5.29}$$

$$\mathcal{F}^{-1}\{F(v)\} = f(t) = \int_{-\infty}^{\infty} F(v)e^{j2\pi vt}\,dv \tag{5.30}$$

Often the angular frequency $\omega = 2\pi v$ is used for (5.30). In this case, $d\omega = 2\pi\,dv$ causes a factor $1/(2\pi)$ in front of (5.30), breaking the symmetric form of the Fourier transform and inverse Fourier transform.

$$F(\omega) = \int_{-\infty}^{\infty} f(t)e^{-j\omega t}\,dt \tag{5.31}$$

$$f(t) = \frac{1}{2\pi} \int_{-\infty}^{\infty} F(\omega)e^{j\omega t}\,d\omega \tag{5.32}$$

To restore the symmetry, sometimes the following definitions are used.

$$F(\omega) = \frac{1}{\sqrt{2\pi}} \int_{-\infty}^{\infty} f(t)e^{-j\omega t}\,dt \tag{5.33}$$

$$f(t) = \frac{1}{\sqrt{2\pi}} \int_{-\infty}^{\infty} F(\omega)e^{j\omega t}\,d\omega \tag{5.34}$$

The expressions (5.29), (5.31) or (5.33) are referred to as Fourier Transform and (5.30), (5.32) or (5.34) are Inverse Fourier Transform.


## 5.4    Some Discussions on Fourier Series

In this section, we look into some details of the Fourier series using examples. Since theFFT algorithm performs Fourier Transform numerically using a finite step for the variable of integration, the discussions made in this section apply to the FFT.

Consider first a straightforward example, i.e., expand a sine function into Fourier series. Figure 5.3a illustrates four periods of a sine function of which amplitude is unity and period is 25 s. The graphs below this time series are the coefficient of the cosine series, of the sine series ($a_k$ and $b_k$ on the right-hand side of (5.1)), the magnitude, and the phase. These graphs are obtained by FFT (Fast Fourier Transform).

We can interpret the meanings of these graphs as follows. Since the given function is a sine function with a period of 25 s, the coefficient of the cosine series is zero over the entire frequency range (Fig. 5.3a2). On the other hand, the coefficient of the sine series shows one peak at 0.04 Hz (Fig. 5.3a3). This frequency is found as the reciprocal of the period 25 s. As mentioned above, the minimum frequency of this case is the reciprocal of the length of the time series, i.e., $1/100 = 0.01$ Hz. Thus, including the DC (0 Hz) term, this only peak appears at the fifth point ($0 + 0.01 \times 4 = 0.04$) on the frequency axis.

The FFT algorithm expands functions using the exponential form of Fourier transform using Euler's notation as discussed in Sect. 5.2. Here, Euler's notation expresses a complex number as $c_n = a_n/2 + j(-b_n/2)$. In the present case, $a_n = 0$ for all $n$'s, and $b_n = 1$ for $n = 4$ (corresponding to 0.04 Hz), and $b_n = 0$ for all the other $n$'s. Therefore, the imaginary part of $c_4$ ($-b_4/2$) is $-1$ as Fig. 5.3a3 indicates.

The above argument expresses the complex number $c_n$ in the form of real and imaginary parts. A complex number can be expressed in the form of magnitude and phase. In the present case, the magnitude is zero at all $n$'s except for $n$ corresponding to 0.04 Hz. At this frequency, since the real part is zero and the imaginary part is $-1$ the magnitude is 1. Figure 5.3a4 indicates this magnitude at 0.04 Hz. Similarly, the phase at this frequency can be evaluated from the arc-tangent of (imaginary part/real part) as $\tan^{-1}((-1)/0) = -\pi/2$. Figure 5.3a5 indicates this phase at 0.04 Hz.

The above-discussed phase of $-\pi/2$ can be argued in a more algebraic way as well. From the mathematical identity $\sin\theta = \cos(\theta - \pi/2)$, we can express the present sine function as follows.

$$\sin\left(2\pi\frac{t}{25}\right) = \cos\left(2\pi\frac{t}{25} - \frac{\pi}{2}\right) \tag{5.35}$$

Equation (5.35) indicates that when expressed as a series of cosine functions, this sine function has a frequency of 0.04 Hz with a phase delay of $-\pi/2$. Figure 5.3a5 indicates this situation by showing that the phase of the FFT at 0.04 Hz is $-\pi/2$.

Figure 5.3b is the case when the sine function in Fig. 5.3a is shifted on the time axis by one-eighth of the period ($\pi/4$ in rad). From the mathematical identity, this function can be expressed as follows.

$$\sin\left(2\pi\frac{t}{25} + \frac{\pi}{4}\right) = \frac{1}{\sqrt{2}}\sin\left(2\pi\frac{t}{25}\right) + \frac{1}{\sqrt{2}}\cos\left(2\pi\frac{t}{25}\right) \tag{5.36}$$

The right-hand side of (5.36) indicates that with the extra phase shift of $\pi/4$, the sine function this time has both the cosine and sine terms oscillating at 0.04 Hz when expanded into the Fourier series. Here, the two terms split the amplitude into two of $1/\sqrt{2} = \sqrt{0.5}$. Figure 5.3b2 and b3 indicate this situation. These two amplitudes make up the total magnitude

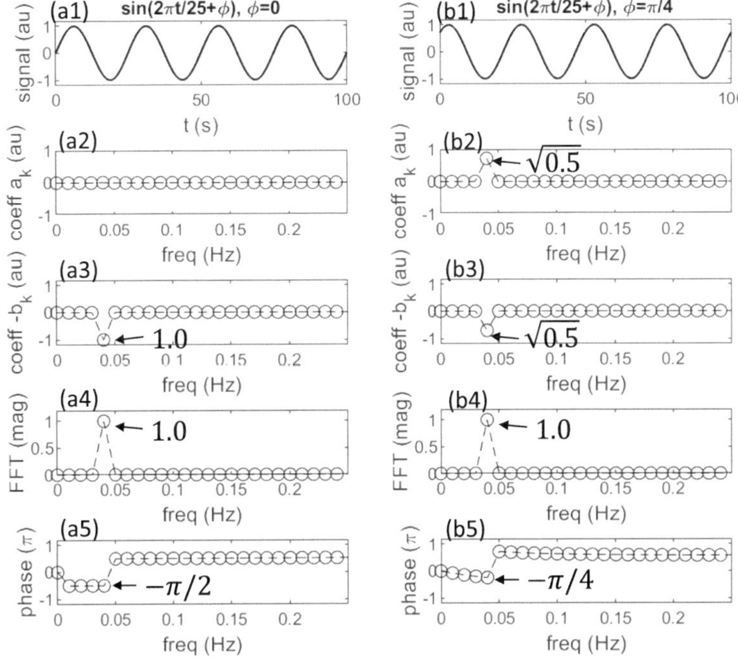


**Fig. 5.3** Sine function with period 25 s. **a1** time series with initial phase $\phi = 0$, **b1** time series with $\phi = \pi/4$; **a2** coefficients $a_k$ for $\phi = 0$, **b2** coefficients $a_k$ for $\phi = \pi/4$; **a3** coefficient $-b_k$ for $\phi = 0$, **b3** coefficient $-b_k$ for $\phi = \pi/4$; **a4** FFT magnitude for $\phi = 0$, **b4** FFT magnitude for $\phi = \pi/4$; **a5** FFT phase for $\phi = 0$, **b5** FFT phase for $\phi = \pi/4$

of unity as $(\sqrt{0.5})^2 + (\sqrt{0.5})^2 = 1$, i.e. the same magnitude as before the phase shift of $\pi/4$, as Fig. 5.3b4 indicates. The magnitude is expected unchanged because a phase change does not alter the magnitude.

Figure 5.3b5 indicates that the phase of the FFT is $\tan^{-1}((-1)/1) = -\pi/4$ at the frequency of 0.04 Hz (arc-tangent of imaginary part/real part). In the more algebraic view as above, we can interpret this phase of $-\pi/4$ as the positive phase shift of $\pi/4$ from (5.35).

$$\sin\left(2\pi\frac{t}{25} + \frac{\pi}{4}\right) = \cos\left(2\pi\frac{t}{25} - \frac{\pi}{2} + \frac{\pi}{4}\right) = \cos\left(2\pi\frac{t}{25} - \frac{\pi}{4}\right) \tag{5.37}$$

Using the coefficients $a_k$ and $b_k$ plotted in Fig. 5.3 for the corresponding frequency, we can reconstruct the signal in the time domain according to (5.1). In this process, often the number of the terms (i.e., the highest $n$ in (5.1)) determines the accuracy of the reconstructed time series. Figure 5.4 compares the reconstructed sine functions used in Fig. 5.3a and b when 10 terms are used (the top graph in Fig. 5.4a and b) and 500 terms are used (the bottom two graphs). In this case, the reconstructed time series does not depend on the number of terms. This is because the original sine function has a single frequency and the fifth frequency bin

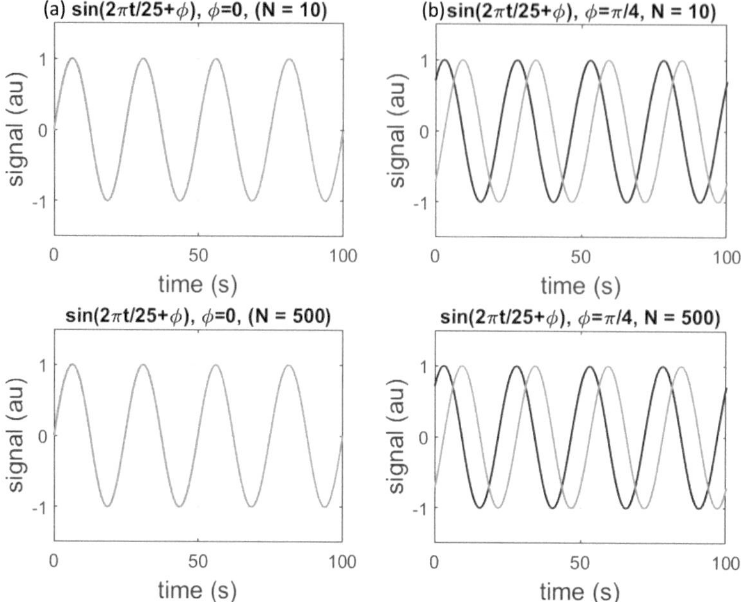



**Fig. 5.4** Reconstructed since wave with period 25 s with initial phase 0 and $\pi/4$

corresponds to this oscillation frequency, as discussed above. Therefore, as long as the fifth terms that correspond to $a_4$ and $b_4$ ($a_0$ is for 0 Hz) are used, the reconstructed time series does not depend on the number of terms. The situation is different when the original function is not a sine or cosine function of a single frequency, as discussed in the following example.

Now consider the square waves function shown in Fig. 5.5. This figure uses the same arrangement as Fig. 5.3. The graphs under the time series at the top are the coefficient of the real part, the coefficient of the imaginary part, the magnitude of the FFT, and the phase of the FFT in this order. The time series on the right has $\pi/4$ phase shift as Fig. 5.3.

Take a quick look at Fig. 5.5a1. From the overall shape and the periodicity of $T = 25$ s, we can speculate that when expanded to a Fourier series this function behaves similar to the sine function we discussed in Fig. 5.3a1. Indeed, Fig. 5.5a2 and a3 exhibit the same feature in Fig. 5.3a2 and a3, i.e., the coefficient of the real part is null and the coefficient of the imaginary part has a peak at 0.04 Hz. The difference from the sine function cases is that the imaginary part coefficients (and the FFT magnitude graph (a4)) have multiple peaks. These peaks indicate that the square wave cannot be expressed with a single sine function, and accordingly, the Fourier series has multiple significant terms.

Figure 5.5b1 and the other graphs under this graph show a similar feature to the case of the sine function shown in Fig. 5.3b1; i.e., the phase shift of $\pi/4$ causes the function to

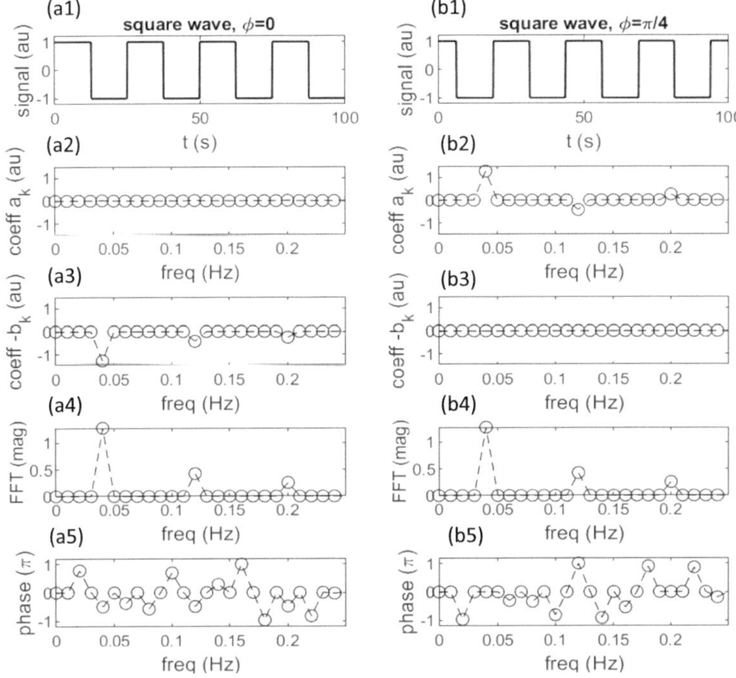


**Fig. 5.5** Periodic square wave function with period $T$. The arrangments of he graphs are the same as Fig. 5.3

behave like cosine function showing peaks in the real part coefficients whereas no peaks in the imaginary part coefficients. The FFT magnitude seen in (b4) appears the same as (a4) as is the case of Fig. 5.3.

Figure 5.6 shows the time series reconstructed based on the coefficients seen in Fig. 5.5 for the square wave cases. Compare these graphs with Fig. 5.4. Unlike the case of the single-frequency sine functions, this time, the accuracy of the reconstructed time series significantly depends on the number of terms used. When we use only 20 terms, the reconstructed time series exhibits ripples. When we use 500 terms, these ripples disappear, and the reconstructed time series overlaps the original square wave functions well.

## 5.5   Some Properties of Fourier Transform

In this section, we briefly discuss several properties of the Fourier transform. More information is available elsewhere [6, 9, 10].

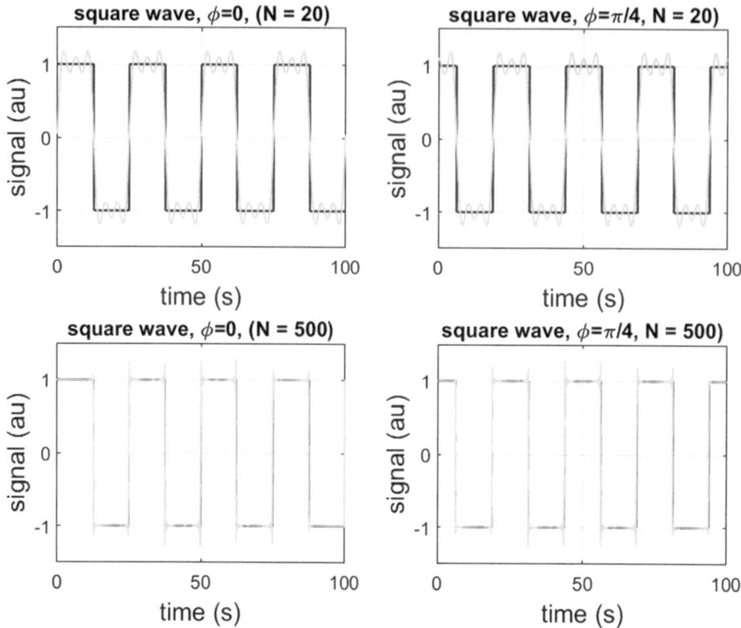


**Fig. 5.6** Reconstructed square waves with period 25 s with initial phase 0 and $\pi/4$

### 5.5.1   Linearity

From (5.29) or (5.31), it is apparent that the Fourier transform preserves the linearity. If function $h(t)$ is a linear combination of $f(t)$ and $g(t)$ as

$$h(t) = af(x) + bg(x) \tag{5.38}$$

so is its Fourier transform.

$$H(\omega) = aF(\omega) + bG(\omega) \tag{5.39}$$

Here the $a$ and $b$ are constant and the upper case denotes the Fourier transform of the lower case.

### 5.5.2   Time Shifting

Shift function $f(t)$ on the time axis by a constant $t_0$ as $f(t - t_0)$ and Fourier transfer the resultant function.

$$\mathcal{F}\{f(t - t_0)\} = \int_{-\infty}^{\infty} f(t - t_0)e^{-j\omega t}dt \tag{5.40}$$

Putting $y = t - t_0$ and noting $dt = dy$, we find

$$\mathcal{F}\{f(t-t_0)\} = \int_{-\infty}^{\infty} f(y)e^{-j\omega(y+t_0)}dt$$

$$= e^{-j\omega t_0} \int_{-\infty}^{\infty} f(y)e^{-j\omega y}dt$$

$$= e^{-j\omega t_0}\mathcal{F}\{f(t)\} = e^{-j\omega t_0}F(\omega) \tag{5.41}$$

Equation (5.41) indicates that shifting by $t_0$ in the time domain corresponds to multiplying by $exp(-j\omega t_0)$ in the frequency domain.

### 5.5.3 Frequency Shifting

Now consider shifting frequency by $\omega_0$ in the frequency domain. From (5.32),

$$\mathcal{F}^{-1}\{F(\omega-\omega_0)\} = \frac{1}{2\pi} \int_{-\infty}^{\infty} F(\omega-\omega_0)e^{j\omega t}d\omega \tag{5.42}$$

Putting $\phi = \omega - \omega_0$ and noting $d\omega = d\phi$, we find

$$\mathcal{F}^{-1}\{F(\omega-\omega_0)\} = \frac{1}{2\pi} \int_{-\infty}^{\infty} F(\phi)e^{j(\phi+\omega_0)t}d\phi$$

$$= e^{j\omega_0 t}\frac{1}{2\pi} \int_{-\infty}^{\infty} F(\phi)e^{j\phi}d\phi$$

$$= e^{j\omega_0 t}\mathcal{F}^{-1}\{F(\phi)\} = e^{j\omega_0 t}f(t) \tag{5.43}$$

Equation (5.43) indicates that shifting by $\omega_0$ in the frequency domain corresponds to multiplying by $exp(j\omega_0 t)$ in the time domain.

### 5.5.4 Time Scaling

Consider scaling the time with a factor $a > 0$. From the definition,

$$\mathcal{F}\{f(at)\} = \int_{-\infty}^{\infty} f(at)e^{-j\omega t}dt$$

$$= \int_{-\infty}^{\infty} f(\tau)e^{-j\omega\frac{\tau}{a}}\frac{d\tau}{a}$$

$$= \frac{1}{a} \int_{-\infty}^{\infty} f(\tau)e^{-j(\frac{\omega}{a})\tau}d\tau$$

$$= \frac{1}{a}F\left(\frac{\omega}{a}\right) \tag{5.44}$$

Here in going from the first to the second line, $at$ is replaced with $\tau$. (5.44) indicates that scaling by a factor $a$ in the time domain corresponds to scaling by $1/a$ the magnitude and frequency in the frequency domain.

## 5.5.5 Frequency Scaling

From (5.44), we readily obtain the following equation.

$$\mathcal{F}^{-1}\{\mathcal{F}\{f(at)\}\} = \frac{1}{a}\mathcal{F}^{-1}\left\{F\left(\frac{\omega}{a}\right)\right\}$$

So,

$$\mathcal{F}^{-1}\left\{F\left(\frac{\omega}{a}\right)\right\} = a\mathcal{F}^{-1}\{\mathcal{F}\{f(at)\}\} = af(at)$$

Defining $b = 1/a$,

$$\mathcal{F}^{-1}\{F(b\omega)\} = \frac{1}{b}f\left(\frac{t}{b}\right) \tag{5.45}$$

Equation (5.45) indicates that scaling by a factor $b$ in the frequency domain corresponds to scaling by $1/b$ the amplitude and time in the time domain.

## 5.5.6 Conjugation

The complex conjugate of the Fourier transform of a function $f(t)$ is related to the Fourier transform of the complex conjugate of $f(t)$ as follows.

$$
\begin{aligned}
[\mathcal{F}\{f(t)\}]^* &= \left[\int_{-\infty}^{\infty} f(t)e^{-j\omega t}dt\right]^* \\
&= \int_{-\infty}^{\infty} [f(t)]^* e^{j\omega t}dt = \int_{-\infty}^{\infty} [f(t)]^* e^{-j(-\omega)t}dt \\
&= \mathcal{F}\{[f(t)]^*\}(-\omega)
\end{aligned}
\tag{5.46}
$$

Here $(\omega)$ represents a function of $\omega$ (not multiplied by $\omega$), e.g., $F\{[f(t)]^*\}(-\omega)$ represents the Fourier transform of $[f(t)]^*$, $F\{[f(t)]^*\}$, as a function of $(-\omega)$. Thus,

$$[\mathcal{F}\{f(t)\}]^*(\omega) = \mathcal{F}\{[f(t)]^*\}(-\omega) \tag{5.47}$$

The Fourier transform of the complex conjugate of a function is equal to the complex conjugate of the Fourier transform of the function with the sign of the frequency flipped.

### 5.5.7 Differentiation

Consider Fourier transform the first order derivative of function $f(t)$. Defining $g(t) \equiv \dot{f}(t)$, we can express the Fourier transform of $g(t)$ as follows.

$$\mathcal{F}\{g(t)\} = \int_{-\infty}^{\infty} \dot{f}(t)e^{-j\omega t}\,dt \tag{5.48}$$

Here $\mathcal{F}\{f(t)\}$ denotes the Fourier transform of function $f(t)$. Integrating the right-hand side by parts, we can rewrite (5.48) as follows.

$$\mathcal{F}\{\dot{f}(t)\} = \left[f(t)e^{-j\omega t}\right]_{-\infty}^{\infty} + j\omega \int_{-\infty}^{\infty} f(t)e^{-j\omega t}\,dt$$

$$= j\omega \int_{-\infty}^{\infty} f(t)e^{-j\omega t}\,dt = j\omega F \tag{5.49}$$

The above expression can be applied to higher-order differentiations. By replacing $g(t)$ with $g(t) = f''(t)$ and expressing the Fourier transform of each function with the upper case letter, we obtain the following expression for the second derivative.

$$\mathcal{F}\{\dot{g}(t)\} = j\omega G \tag{5.50}$$

From (5.49), $G = \mathcal{F}\{\dot{f}(t)\} = j\omega F$. Substituting this into (5.50), we find as follows.

$$\mathcal{F}\{\ddot{f}(t)\} = \mathcal{F}\{\dot{g}(t)\} = j\omega G = j\omega(j\omega F) = (j\omega)^2 F \tag{5.51}$$

In this fashion, we can express the Fourier transform of derivatives of $n$th order in the form of $(j\omega)^n F$.

### 5.5.8 Integration

Repeating the same argument as the differentiation case, we can derive the following relation for integration. Defining $g(t) \equiv \int f(t)dt$, we can express the Fourier transform of $g(t)$ as follows.

$$\mathcal{F}\{g(t)\} = \int_{-\infty}^{\infty} \left(\int f(t)dt\right) e^{-j\omega t}\,dt$$

$$= \frac{-1}{j\omega}\left[\left(\int f(t)dt\right)e^{-j\omega t}\right]_{-\infty}^{\infty} + \frac{1}{j\omega}\int_{-\infty}^{\infty} f(t)e^{-j\omega t}\,dt \tag{5.52}$$

$$= \frac{1}{j\omega}\mathcal{F}\{f(t)\}$$

Thus,

$$\mathcal{F}\left\{\int f(t)dt\right\} = \frac{1}{j\omega}\mathcal{F}\{f(t)\} \tag{5.53}$$

## References


1. James JF (2007) A student's guide to Fourier transforms: with applications in physics and engineering, 3rd edn. Cambridge University Press, Cambridge
2. Wiener N (1933) The Fourier integral and certain of its applications. Cambridge University Press, New York, USA (Reissued in digital print 2008)
3. Boas ML (2006) Mathematical methods in the physical sciences, 3rd edn. Wiley, London. Chap. 7
4. Davis HF (1989) Fourier series and orthogonal functions, 1st edn. Dover, New York, USA
5. Bloomfield P (2000) Fourier analysis of time series: an introduction, 2nd edn. Wiley-Interscience, New York
6. Smith JO III (2008) Mathematics of the discrete Fourier transform (DFT): with audio applications, 2nd edn. W3K Publishing. http://books.w3k.org/
7. Nyquist H (1928) Certain topics in telegraph transmission theory. Trans AIEE 47(2):617–644
8. Poberezhskiy YS, Gennady P, Poberezhskiy G (2019) Signal digitization and reconstruction in digital radios. Artech House, Norwood, MA, USA
9. Hansen EW (2014) Fourier transforms. Principles and applications. Wiley, Hoboken, New Jersey
10. Mavko G, Mukerji T, Dvorkin J (2020) The rock physics handbook, 3rd edn. Cambridge University Press, New York. Chap. 1

# Linear Systems and Frequency Domain Analysis 6

A linear system is a system that responds to input linearly. A spring-point-mass system is an example because the point-mass's displacement (output) is proportional to the applied force (input). We can also say that the equation of motion that governs the dynamics of a linear system is a linear differential equation. The coefficient of all the terms is constant.

The Laplace Transform is a convenient tool to describe the dynamics of linear systems in the frequency domain. Using the Laplace Transform, we can solve differential equations algebraically. In this chapter, after briefly reviewing linear systems, we discuss the basic properties of the Laplace Transform. Then, we use the Laplace Transform to solve the differential equation for typical physical systems.

## 6.1 Linear Systems

A linear system [1] is defined as a system consisting of a linear operator. A linear operator operates on quantities as input and outputs a quantity where the output is a linear combination of input quantities. Mathematically the linearity is defined as follows.

$$H\{x_1\} = y_1 \tag{6.1}$$

$$H\{x_2\} = y_2 \tag{6.2}$$

$$H\{x_1 + x_2\} = y_1 + y_2 \tag{6.3}$$

Here $H$ represents a linear operator that takes $x_i$ as input and outputs $y_i$. The input can be a function of an independent variable such as time. This function does not have to be linear. In science and engineering, a sinusoidal function is often input to a linear system (operator), as will be discussed in the remainder of this section.

© The Author(s), under exclusive license to Springer Nature Switzerland AG 2025    91
S. Yoshida, *Physics and Mathematics Behind Wave Dynamics*, Synthesis Lectures on Wave Phenomena in the Physical Sciences,
https://doi.org/10.1007/978-3-031-60354-9_6

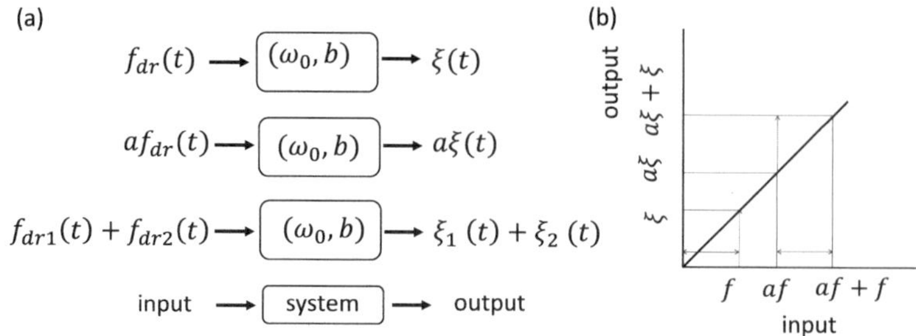


**Fig. 6.1** Conceptual diagram of a linear system. Suffix '*dr*' stands for 'driving'

Figure 6.1a illustrates the concept of linear systems schematically. The input is a function of time and it drives the linear system. The linear system is characterized by natural frequency $\omega_0$ and damping coefficient $b$. (See Chap. 2.) When the input signal becomes greater by a factor of $a$, so does the output. If the input is a superposition of two signals, the output is the superposition of the output of the individual input. Figure 6.1b exhibits this property by using $f(t)$, $af(t)$ and $f(t) + af(t)$ as input. Notice that the output of $af(t)$, $a\xi(t)$, is greater than the output of $f(t)$, $\xi(t)$, by a factor of $a$. The output of $f(t) + af(t)$ is the addition of $\xi(t)$ and $a\xi(t)$. These linear relations hold at any moment $t$.

In the following sections, we discuss various mathematical concepts applicable to linear systems.

## 6.2    Laplace Transform

The Laplace transform [2–4] of the function $f(t)$ is defined as follows.

$$\mathcal{L}\{f(t)\} = \int_0^\infty f(t)e^{-st}dt \tag{6.4}$$

where $s$ is a complex number ($\sigma$ and $\omega$ are real).

$$s = \sigma + j\omega \tag{6.5}$$

### 6.2.1    Laplace Transform of Several Functions

**Unit step (Heaviside) function** $E(t)$

$$\mathcal{L}\{E(t)\} = \int_0^\infty e^{-st}dt = \left[\frac{-1}{s}e^{-st}\right]_0^\infty = \frac{1}{s} \tag{6.6}$$

**Delta function** $\delta(t)$

$$\mathcal{L}\{\delta(t)\} = \int_0^\infty \frac{dE(t)}{dt} e^{-st} dt = \left[E(t)e^{-st}\right]_0^\infty - \int_0^\infty E(t)(-s)e^{-st} dt$$

$$= 0 + s \int_0^\infty e^{-st} dt = s\frac{1}{s} = 1 \tag{6.7}$$

**Sine function** $\sin at$

$$\mathcal{L}\{\sin at\} = \int_0^\infty \sin at\, e^{-st} dt = \left[\frac{-1}{s}e^{-st} \sin at\right]_0^\infty + \frac{a}{s}\int_0^\infty \cos at\, e^{-st} dt$$

$$= 0 + \frac{a}{s}\left\{\left[\frac{-1}{s}e^{-st} \cos at\right]_0^\infty - \frac{a}{s}\int_0^\infty \sin at\, e^{-st} dt\right\}$$

$$= \frac{a}{s^2} - \frac{a^2}{s^2}\mathcal{L}\{\sin at\}$$

Therefore,

$$\mathcal{L}\{\sin at\} = \frac{a}{s^2 + a^2} \tag{6.8}$$

**Cosine function** $\cos at$

$$\mathcal{L}\{\cos at\} = \int_0^\infty \cos at\, e^{-st} dt = \left[\frac{-1}{s}e^{-st} \cos at\right]_0^\infty - \frac{a}{s}\int_0^\infty \sin at\, e^{-st} dt$$

$$= \frac{1}{s} - \frac{a}{s}\left\{\left[\frac{-1}{s}e^{-st} \sin at\right]_0^\infty + \frac{a}{s}\int_0^\infty \cos at\, e^{-st} dt\right\} \tag{6.9}$$

$$= \frac{1}{s} - \frac{a^2}{s^2}\mathcal{L}\{\cos at\}$$

Therefore,

$$\mathcal{L}\{\cos at\} = \frac{s}{s^2 + a^2} \tag{6.10}$$

### 6.2.2 Some Properties of Laplace Transform

Here we discuss some properties of Laplace Transform. More information is available elsewhere [2, 4, 5].

**Linearity**

From definition (6.4), the Laplace transform of the addition of functions is the addition of the Laplace transforms of the individual function.

$$\mathcal{L}\{f(t) + g(t)\} = \int_0^\infty \{f(t) + g(t)\} e^{-st} dt$$

$$= \int_0^\infty f(t)e^{-st} dt + \int_0^\infty g(t)e^{-st} dt = \mathcal{L}\{f(t)\} + \mathcal{L}\{g(t)\} \tag{6.11}$$

Similarly, the following relation holds.

$$\mathcal{L}\{af(t)\} = a \int_0^\infty f(t)e^{-st} dt = a\mathcal{L}\{f(t)\} \tag{6.12}$$

**Differentiation in time domain**

$$\mathcal{L}\{f'(t)\} = \int_0^\infty f'e^{-st} dt = \left[ fe^{-st} \right]_0^\infty + s \int_0^\infty fe^{-st} dt$$

$$= -f(0) + s\mathcal{L}\{f(t)\} \tag{6.13}$$

**Integration in time domain**

$$\mathcal{L}\left\{ \int_0^t f(t)dt \right\} = \int_0^\infty \left( \int_0^t f(t)dt \right) e^{-st} dt$$

$$= -\frac{1}{s} \left[ \left( \int_0^t f(t)dt \right) e^{-st} \right]_0^\infty + \frac{1}{s} \int_0^\infty f(t)e^{-st} dt$$

$$= -\frac{1}{s} \left[ \left\{ \left( \int_0^\infty f(t)dt \right) e^{-\infty} \right\} - \left\{ \left( \int_0^0 f(t)dt \right) e^{-0} \right\} \right] + \frac{1}{s}\mathcal{F}\{f(t)\}$$

$$= \frac{1}{s}\mathcal{F}\{f(t)\} \tag{6.14}$$

## 6.3    Convolution Integral

Response of a linear system can be described by the operation known as the convolution integral [2, 6, 7], as shown in the following expression.

$$y(t) = \int_0^t x(\tau)w(t - \tau)d\tau \tag{6.15}$$

Here $x(t)$ is the input to the linear system, $y(t)$ is the output (response) of the system, and $w(t - \tau)$ represents the linear system. The function $w(t)$ characterizes how the linear system transfers the operation defined by the input. From this standpoint, often $w(t)$ is referred to as the transfer function [8, 9] of the linear system. As will be discussed shortly, the transfer function is the impulse response of the system as well. This indicates that if we know the impulse response of a linear system, we can fully characterize the system. In this section, we discuss this in various ways.

Take some time to consider a linear system and its impulse response. A linear system is a system in which the output is linearly related to the input. One of the simplest examples of a linear system is a spring-point-mass system. Here, the linearity arises from Hooke's law, which describes the linear relationship between the applied force, $f(t)$, and the resulting displacement from the equilibrium, $\xi(t)$.

$$f(t) = -k\xi(t) \tag{6.16}$$

Here the constant of proportionality $k$ is known as the spring constant. The negative sign in front of $k$ indicates that the displacement vector is always opposite to the applied force vector.

In more mathematical terms, the linearity associated with a spring-point-mass system can be expressed as follows. Imagine that two external agents apply force to the mass. Call the two forces $f_1$ and $f_2$. At each moment, the system responds to the forces causing displacement $\xi_1$ and $\xi_2$. Here the force (input) and the displacement (output) are related through the spring constant $k$ as $f_1 = -k\xi_1$ and $f_2 = -k\xi_2$. The total input $f_1 + f_2$ is related to the total displacement as $f_1 + f_2 k = -k(\xi_1 + \xi_2)$. Of course, if the input is increased by a factor of $a$ as $a(f_1 + f_2)$, so is the output as $-a/k(\xi_1 + \xi_2)$. The input and output are related linearly.

Consider that a certain force is applied as a function of time $f(t)$. At each moment, the external agent exerts force with a certain magnitude. The applied force can be expressed as a train of impulse functions (unit impulse functions weighted by the function $f(t)$) along the time axis weighted by the function $f(t)$. Thus, the effect of this force can be viewed as a series of impulses occurring continuously as $f(t) = f(\tau)\delta(t - \tau)$. Here $\delta(\tau)$ is the unit impulse function and $t$ represents the delay. Figure 6.2 schematically illustrates the situation where the input can be expressed as a continuous series of impulses where the amplitude at time $t$ varies as described by the function $f(t)$. The system responds to the impulse input at each moment linearly and the resultant displacement (output) is the superposition of the current impulse response and the cumulative response (the hysteresis) of the linear system.

Formulaically, the above situation can be expressed as follows. Referring to Fig. 6.2 consider the moment represented by time $t_i$. At this moment, the input to the system can be expressed as

$$x_i = f(t_i) \tag{6.17}$$

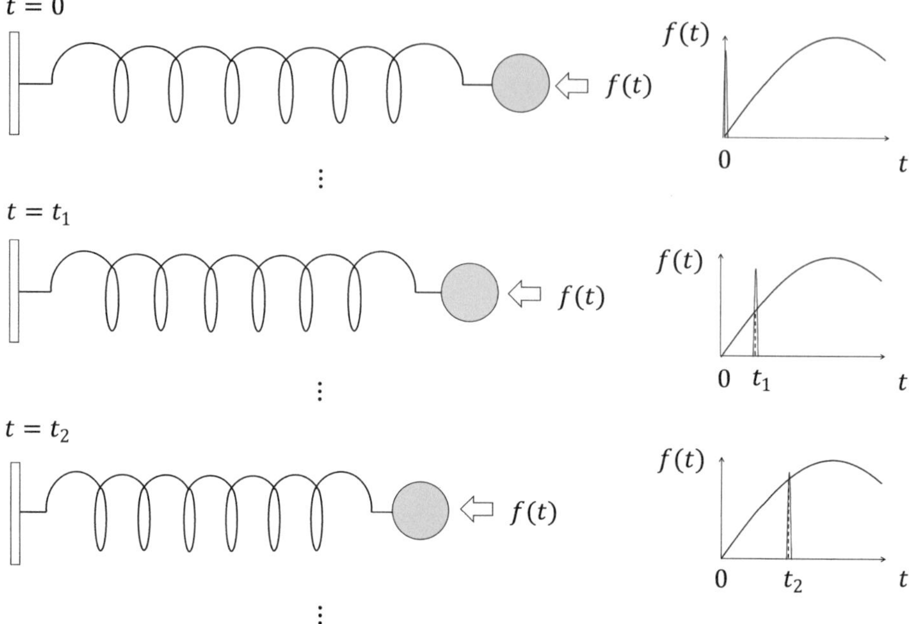

**Fig. 6.2** Linear system responds to input as a continuous series of impulse input

The entire input function defined in $0 \le t \le \tau$ can be expressed as follows.

$$x(\tau) = \int_0^\tau f(\xi)\delta(\tau - \xi)d\xi \tag{6.18}$$

Now consider the output, i.e., the response of the system to the input. As mentioned above, the output at a given time $t_{now}$ is the superposition of the at-this-moment impulse response to the input $f(t_{now})$ and the hysteresis due to the input in the past $(0 \le t < t_{now})$. Assume that the system has non-zero input only at $t = t_1$, $t = t_3$, and $t = t_{now}$ $(t_1 < t_3 < t_{now})$. The output at $t = t_{now}$ is the superposition of the effects at $t = t_1$, $t = t_3$, and $t = t_{now}$, and can be expressed as follows

$$y(t_{now}) = x(t_1)w(t_{now} - t_1) + 0 + x(t_3)w(t_{now} - t_3) + 0 + x(t_{now})w(t_{now} - t_{now}) \tag{6.19}$$

Here $w(t)$ is the response function. The quantity $(t_{now} - t_i)$ represents the time elapse from the moment when the system received the impulse input. Assume that the response is monotonic decay. At time $t_{now}$ the effect due to the input at $t_1$, $x(t_1)$, has decayed according to $w(t_{now} - t_1)$, the effect due to $x(t_3)$ has decayed as $w(t_{now} - t_3)$, and the effect due to the at-this-moment $x(t_{now})$ has not decayed because $w(t_{now} - t_{now}) = w(0)$ represents the impulse response now (before any decay). (Imagine you injured your right knee three years ago and now reinjured. You have almost recovered from the injury of three years ago but still feel discomfort. The new injury you just experienced adds a severe pain to this discomfort.)

Next consider that the input is continuous. For example, you tap the point-mass connected to a spring with a constant interval of $\Delta t$ starting at $t = 0$. The output at $t = 3\Delta t$ (current time or $t_{now}$) can be expressed as follows.

$$y(3\Delta t) = x(0)w(3\Delta t - 0) + x(\Delta t)w(3\Delta t - \Delta t) + x(2\Delta t)w(3\Delta t - 2\Delta t)$$
$$+ x(3\Delta t)w(3\Delta t - 3\Delta t)$$
$$= x(0)w(3\Delta t) + x(\Delta t)w(2\Delta t) + x(2\Delta t)w(\Delta t) + x(3\Delta t)w(0) \qquad (6.20)$$

Here the last term on the right-hand side of (6.20) represents the output due to current input, the second from the last term represents the residual effect due to the input $\Delta t$ s ago, the third term represents the residual effect due to the input $2\Delta t$ s ago, etc. You can easily imagine that the point-mass moves (perhaps oscillates) due to the initial tap at $t = 0$, when you apply the second tap at $t = \Delta t$. Similarly, at a given tap the point-mass has residual motion due to the past tapping inputs.

By making the time step infinitesimally small, we can express (6.20) in the following form.

$$y(t) = \int_0^t x(\tau)w(t - \tau)d\tau \qquad (6.21)$$

Substituting (6.18) into (6.21), we obtain the following expression for the output as a function of time.

$$y(t) = \int_0^t x(\tau)w(t - \tau)d\tau = \int_0^t \left( \int_0^\tau f(\xi)\delta(\tau - \xi)d\xi \right) w(t - \tau)d\tau$$
$$= \int_0^t f(\xi) \left( \int_0^\tau \delta(\tau - \xi)w(t - \tau)d\tau \right) d\xi \qquad (6.22)$$

Since the integrand of the second integration inside the parenthesis is non-zero only when $\tau = \xi$, this integration can be replaced by $w(t - \xi)$. Thus, we can express the output $y(t)$ using the applied force and the function $w(t)$ as follows.

$$y(t) = \int_0^t f(\xi)w(t - \xi)d\xi \qquad (6.23)$$

Expression (6.23) is known as the convolution integral.

It is possible to view this equation as it connects the input to the output through function $w(t)$. In other words, we can interpret that (6.23) expresses how the applied force as the input to the linear system is "transferred" to the output of the system. If the linear system is a spring-point-mass system and the output $y(t)$ is the displacement of the mass, $w(t)$ represents this transferring effect from force to displacement in the time domain. The corresponding function in the frequency domain is the transfer function, as will be discussed shortly.

The above observation indicates that the transfer function is related to the convolution integral. Indeed, the two concepts are strongly related to each other. This relation is somewhat hard to understand intuitively in the time domain. However, in the frequency domain, it is much more easily understood. The discussion of the transfer function in conjunction with the convolution integral is the main topic of the following section.

Before continuing the discussion it is useful to consider (6.23) under the condition that the applied force is an impulse. Setting $f(t) = f_0 \delta(t)$, we can express this situation as follows.

$$y(t) = f_0 \int_0^t \delta(\xi)w(t - \xi)d\xi = f_0 w(t) \tag{6.24}$$

Equation 6.24 indicates that the transfer function can be interpreted as the impulse response of the linear system. This interpretation helps us view the convolution integral (6.23) as representing the consecutive impulse response of the system. This topic will be revisited below under 6.5 "Some arguments on the transfer function and convolution".

### 6.3.1   Convolution in Laplace (Frequency) Domain

The convolution integral discussed in the preceding section can be discussed elegantly in the frequency domain using the Laplace transform. Consider the Laplace transform of convolution using two functions $f(t)$ and $g(t)$.

$$\mathcal{L}\{f * g\} = \int_0^\infty \left( \int_0^t f(\xi)g(t - \xi)d\xi \right) e^{-st} dt$$

Here "*" represents the convolution in the form of (6.23). Putting $\tau = t - \xi$

$$\begin{aligned}
\mathcal{L}\{f * g\} &= \int_0^\infty \left( \int_0^t f(\xi)g(\tau)d\xi \right) e^{-s(\tau+\xi)} d\tau \\
&= \left( \int_0^\infty g(\tau)e^{-s\tau} d\tau \right) \left( \int_0^\infty f(\xi)e^{-s\xi} d\xi \right) \\
&= \mathcal{L}\{f\} \mathcal{L}\{g\} \tag{6.25}
\end{aligned}$$

Note that the upper limit of integration with respect to $\xi$ has been replaced with $\infty$. We can do so because $\xi > t$ represents the future, and therefore function $f(\xi)$ has no value; there is no future input. In other words, increasing the upper limit from t to $\infty$ has no effect.

Equation (6.25) indicates that in the Laplace (frequency) domain, the convolution of two functions takes the form of the product of the Laplace transforms of the respective functions.

Apply this Laplace domain expression of the convolution to (6.23) in the context of the output of a linear system as the convolution integral of the input and transfer function. We can immediately obtain the following expression.

$$\mathcal{L}\{y\} = \mathcal{L}\{x\}\,\mathcal{L}\{w\} \tag{6.26}$$

Equation (6.26) states that we can find the Laplace transform of the output of a linear system by multiplying the Laplace transform of the input by the Laplace transform of the transfer function. Using the fact that the transfer function is the impulse response of the linear system, we can rephrase this statement as the Laplace transform of the output of a linear system is the product of the Laplace transform of the input and the Laplace transform of the impulse response of the linear system.

## 6.4  Laplace Transform and Fourier Transform

Compare the exponential form of Fourier transform defined by (5.31) and the Laplace transform (6.4) with the definition of $s = \sigma + j\omega$ (6.5).

$$\mathcal{F}\{f(t)\} = F(\omega) = \int_{-\infty}^{\infty} f(t)e^{-j\omega t}\,dt \tag{5.31}$$

$$\mathcal{L}\{f(t)\} = \int_{0}^{\infty} f(t)e^{-st}\,dt = \int_{0}^{\infty} f(t)e^{-(\sigma+j\omega)t}\,dt \tag{6.4}$$

Comparing the right-hand sides of expressions (5.31) and (6.4), we can interpret that the Fourier transform and Laplace transform are equivalent when the source function $f(t)$ is defined only for $t > 0$ and has negligibly small decay $\sigma << 1$. In most engineering applications, we define the function for $t > 0$. Also, it is not unusual that the decaying characteristic of the function is negligible. In these cases, we can use the Fast Fourier Transform algorithm for various operations in which the Laplace Transform is applied.

As far as the convolution integral is concerned, the Fourier transform behaves the same as the Laplace transform. Consider the Fourier transform of the convolution integral in the context of the response of a linear system. From (6.23) and the definition of the Fourier transform (5.31), we obtain the following expression.

$$\mathcal{F}\{y(t)\} = \int_{-\infty}^{\infty}\left[\int_{-\infty}^{\infty} x(\tau)w(t-\tau)d\tau\right]e^{-j\omega t}\,dt$$

Putting $\eta = t - \tau$

$$\mathcal{F}\{y(t)\} = \int_{-\infty}^{\infty}\left[\int_{-\infty}^{\infty} x(\tau)w(\eta)d\tau\right]e^{-j\omega(\eta+\tau)}\,d\eta$$

$$= \int_{-\infty}^{\infty} x(\tau)e^{-j\omega\tau}\,d\tau \int_{-\infty}^{\infty} w(\eta)e^{-j\omega\eta}\,d\eta$$

$$= \mathcal{F}\{x(t)\}\mathcal{F}\{w(t)\} \tag{6.27}$$

Here $x(t)$ is the input to a linear system and $w(t)$ is the transfer function of the system.

Equation (6.27) indicates that in the frequency domain, the convolution integral becomes the product of the Fourier transforms of the input and the transfer function. Using the upper case symbol to represent the Fourier transform, we can put (6.27) in the following form.

$$Y(\omega) = X(\omega)W(\omega) \tag{6.28}$$

Alternatively,

$$W(\omega) = \frac{Y(\omega)}{X(\omega)} \tag{6.29}$$

In Sect. B.1 we will find that the Fourier transform of an impulse signal is unity (flat in the frequency domain). So, when the input $x(t)$ is an impulse, (6.28) becomes as follows.

$$Y_\delta(\omega) = X_\delta(\omega)W(\omega) = E(\omega)W(\omega) = W(\omega) \tag{6.30}$$

Here $E(\omega)$ is the unit step function defined on the frequency axis. Equation (6.30) indicates that the Fourier transform of the impulse response is the transfer function from the input to the output. This means that we can find the transfer function of a linear system by feeding an impulse input and examining the output. In a later section, we will derive the transfer function of a spring-mass system in a different manner.

The fact that in the frequency domain, the output of a linear system can be expressed as the product of the Fourier transform of the input and the transfer function is very convenient when the output of a linear system is the input to another linear system. If several linear systems are connected in series, the overall transfer function can be expressed as the product of the Fourier transform of the respective systems. For instance, consider that you are designing a comfortable suspension system for a passenger car. The input is the shocks (disturbance) that the tires feel depending on the road surface. The frequency of the disturbance depends on the speed of the car and the road condition. This disturbance is transferred to the seats through the air in the tire, spring mechanism, shock absorbers, and the mechanism that connects the seat to its wheels. The overall transfer function is the product of each of these components. You can design each component separately, find the Fourier transform, multiply them, and evaluate the product.

## 6.5   Some Arguments on the Transfer Function and Convolution

### 6.5.1   Transfer Function as Impulse Response

The above argument indicates that the output signal $y(t)$ of the linear system represented by $w(t)$ for an input $x(t)$ can be expressed as the convolution integral as follows.

$$y(t) = \int_{-\infty}^{\infty} w(t - \tau)x(\tau)d\tau = \int_{0}^{\infty} w(t - \tau)x(\tau)d\tau \tag{6.31}$$

Consider the Laplace transform of $y(t)$.

$$\mathcal{L}\{y(t)\} = \int_0^\infty \int_0^\infty w(t-\tau)x(\tau)d\tau e^{-st}dt$$

$$= \int_0^\infty \int_0^\infty w(\xi)x(\tau)d\tau e^{-s(\xi+\tau)}d\xi$$

$$= \left[\int_0^\infty w(\xi)e^{-s\xi}d\xi\right]\left[\int_0^\infty x(\tau)e^{-s\tau}d\tau\right]$$

$$= \mathcal{L}\{w(t)\}\,\mathcal{L}\{x(t)\} \tag{6.32}$$

The transfer function is defined as

$$W(s) \equiv \mathcal{L}\{w(t)\} = \frac{\mathcal{L}\{y(t)\}}{\mathcal{L}\{x(t)\}} \equiv \frac{Y(s)}{X(s)} \tag{6.33}$$

Consider an impulse input as a special case. From the discussion in Sect. 6.2, we know that the Laplace transform of the delta function is $W(s) = 1$. So, in this special case (6.33) becomes as follows.

$$W(s) = \frac{\mathcal{L}\{y_\delta(t)\}}{\mathcal{L}\{\delta(t)\}} = \frac{Y_\delta(s)}{1} = Y_\delta(s) \tag{6.34}$$

Equation (6.34) indicates that the transfer function is the impulse response of the linear system.

We can say that in the time domain, the input-output relation of the linear system is represented by the convolution integral (6.31) via the time-domain expression of the weight function $w(t-\tau)$, and that in the Laplace (frequency) domain, the input-output relation of the linear system is represented by (6.33) via the transfer function $W(s)$ which is the impulse response of the linear system.

### 6.5.2  Duhamel's Principle

It is interesting to make the following argument. Consider external force $f(t)$ acts on a linear mechanical system. In response to $f(t)$, the mechanical system exhibits some dynamic behavior represented by $\xi(t)$ as schematically illustrated by Fig. 6.3. We can view $f(t)$ as input to the linear system and $\xi(t)$ as the output. Here $\xi(t)$ can be displacement, velocity, or other quantity depending on how you define the linear system. In any event, we can find the output $\xi(t)$ as a solution to a linear differential equation, which can be expressed in general in the following form.

$$a_n\xi^{(n)} + a_{n-1}\xi^{(n-1)} + \cdots + a_1\xi^{(1)} + a_0\xi = f(t) \tag{6.35}$$

$$x^{(n-1)}(0) = \xi^{(n-2)}(0) = \cdots = \xi^{(1)}(0) = \xi(0) = 0 \tag{6.36}$$

Here $\xi^{(n)}$ represents the $n^{th}$ order differentiation of $\xi(t)$ with respect to time $t$, $a_n \cdots a_0$ are all constants, and (6.36) indicates that the initial values are all zero meaning that the system is completely stationary until the external force starts acting on it.

From (6.31) we can relate $x(t)$ and $f(t)$ as follows.

$$\xi(t) = \int_0^t w(t - \tau)f(\tau)d\tau \tag{6.37}$$

$$\int_0^t h(t - \tau)\xi(\tau)d\tau = f(t) \tag{6.38}$$

Here $w(t - \tau)$ represents the linear system where $f(\tau)$ is the input and $\xi(t)$ is the output, and $h(t - \tau)$ represents the linear system where $x(\tau)$ is the input and $f(t)$ is the output.

Now consider the above expressions in the Laplace domain. Express the Laplace transform of the above time domain functions with the uppercase. From (6.13) and the initial conditions (6.36), the differential equation (6.35) can be written as follows.

$$\left(a_n s^{(n)} + a_{n-1}s^{(n-1)} + \cdots + a_1 s^{(1)} + a_0 s\right) \Xi(s) = F(s) \tag{6.39}$$

Here $\Xi(s)$ and $F(s)$ are the Laplace transform of $\xi(t)$ and $f(t)$.

As we discussed in Sect. 6.3, the convolution in the time domain becomes the product in the Laplace domain. Thus, we obtain the following equations from (6.37) and (6.38).

$$\Xi(s) = W(s)F(s) \tag{6.40}$$

$$F(s) = H(s)\Xi(s) \tag{6.41}$$

Equations (6.40) and (6.41) indicate that $W(s)$ is the transfer function from the input to the output of the linear system, and $H(s)$ is the transfer function from the output to the input of the same linear system. Further, a comparison of (6.39) and (6.41) indicates that the differential equation that represents the linear system expresses the output to input transfer function $H(s)$. Summarizing all these arguments, we obtain the following relations.

$$W(s) = \frac{1}{H(s)} = \frac{1}{a_n s^{(n)} + a_{n-1}s^{(n-1)} + \cdots + a_1 s^{(1)} + a_0 s} \tag{6.42}$$

Equations (6.40) and (6.42) tell us that when an input is given to the linear system we can express the output in the Laplace domain with the use of the transfer function $W(s)$, which in turn, can be expressed with the linear coefficient of the differential equation that represents the dynamics.

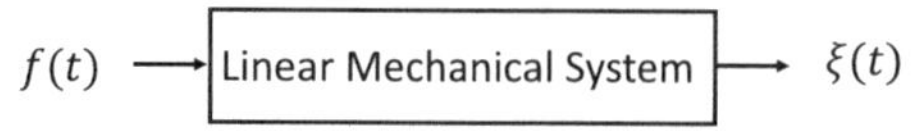

$$f(t) \longrightarrow \boxed{\text{Linear Mechanical System}} \longrightarrow \xi(t)$$

**Fig. 6.3** Linear mechanical system outputs $\xi(t)$ in response to input $f(t)$

The integral form of the input-to-output transfer function $W(s)$ is useful to discuss some physical concepts underlying the above arguments. Let $K(s)$ be the integral form of $W(s)$.

$$W(s) = sK(s) \tag{6.43}$$

Equation (6.43) indicates that differentiation of $K(s)$ is $W(s)$, hence $K(s)$ is the integral form of the transfer function representing the linear system. Express (6.40) with $K$ as follows.

$$\Xi(s) = sKF = sKF + f_0 K - f_0 K = f_0 K + (-f_0 + sF)K \tag{6.44}$$

Here for simplicity, $(s)$ is omitted in the Laplace transforms. Remembering that the time domain differentiation of function $f$ can be expressed in the Laplace domain with the Laplace transform $F$ and the initial condition (6.13), we can interpret that the expression inside the parenthesis in (6.44) as the Laplace domain expression of the derivative of $f(t)$. Hence, the second term on the right-hand side is the product of the Laplace transform of the derivative and function $k(t)$. Thus, the time domain expression of (6.44) can be written as follows.

$$\xi(t) = f(0)k(t) + \int_0^t f'(t - \tau)k(\tau)d\tau \tag{6.45}$$

Expression (6.45) is known as Duhamel's integral (Duhamel's theorem). [10, 11].

Now consider the physical meaning of $k(t)$ as the integral form of $w(t)$ in the time domain.

$$k(t) = \int_0^t w(t)dt \tag{6.46}$$

Equation (6.34) indicates that the transfer function is the impulse response of the system. So, $k(t)$ can be interpreted as the integration of the impulse response over time. In other words, $k(t)$ represents the output of the system when impulse input is continuously applied. Such an input can be naturally expressed as the time integral of the delta function.

$$E(t) = \int_0^t \delta(t)dt \tag{6.47}$$

Here $E(t)$ is the unit step (Heaviside) function. So, we can call $k(t)$ as the step response (normal response) of the system. From all these arguments, we can interpret (6.45) as expressing that the output of the linear system is an accumulation of the normal response. The first term of this equation represents the system's normal response to the initial value of the input $f(0)$. The second integral represents the cumulative effect of the addition of the normal response for $df = f'd\tau$ occurring at each time segment $d\tau$ as time goes by. Figure 6.4 illustrates this schematically. The output $\xi(t)$ can be evaluated as the area of this graph from $\tau = 0$ to $\tau = t$. If the input is constant ($f' = 0$), the output becomes simply the initial value $f(0)$ times the duration $t$.

**Fig. 6.4** Physical meaning of expression (6.45)

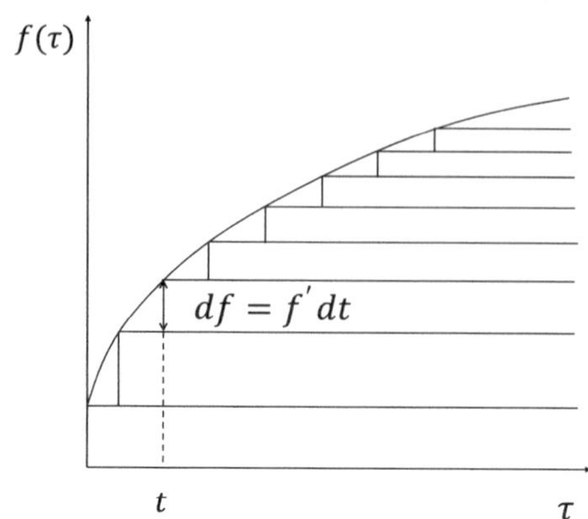


### 6.5.3   Simple Example of Linear System's Output as Convolution Integral

Consider a spring-point-mass system as an example of a linear system. The mass is $m$ and the spring constant is $k$. The equation of motion of this system can be expressed by the following linear differential equation.

$$\ddot{\xi}(t) + \omega_0^2 \xi(t) = \alpha \sin \Omega t \tag{6.48}$$

Here $\omega_0 = \sqrt{k/m}$ is the natural angular frequency of the system and $\alpha = F/m$ is the acceleration due to the external force of which amplitude is $F$ and frequency is $\Omega$.

First, solve the differential equation (6.48) in the time domain putting the general solution as the summation of the homogeneous solution $\tilde{\xi}$ and the particular solution $\bar{\xi}$. The following form of $\tilde{\xi}$ satisfies the differential equation (6.48) when the right-hand side (the source term) is zero.

$$\tilde{\xi}(t) = a \sin(\omega_0 t + \phi) \tag{6.49}$$

Here the amplitude $a$ and the initial phase $\phi$ are determined by the initial condition. For simplicity, set $\phi = 0$ in the following discussion. This setting does not cause us to lose the gist of our discussion here.

The particular solution can take the following form.

$$\bar{\xi}(t) = A \sin \Omega t \tag{6.50}$$

Substitute solution (6.50) into differential equation (6.48) and obtain the following equation.

$$-\Omega^2 A \sin \Omega t + \omega_0^2 A \sin \Omega t = \alpha \sin \Omega t \tag{6.51}$$

This determines the amplitude $A$ as

$$A = \frac{\alpha}{\omega_0^2 - \Omega^2} \tag{6.52}$$

and hence allows us to express the general solution as follows.

$$\xi(t) = \tilde{\xi}(t) + \bar{\xi}(t) = a \sin \omega_0 t + \frac{\alpha}{\omega_0^2 - \Omega^2} \sin \Omega t \tag{6.53}$$

Now solve the differential equation (6.48) using the convolution. According to (6.21), we should be able to find output $y(t)$ from the convolution integral of the impulse response $w(t - \xi)$ and the input function $f(\xi)$. In the differential equation (6.48) the source term $\sin \Omega t$ can be interpreted as the input to the system. Let $y(t)$ be the output and express it based on (6.21). This output should represent the solution to (6.48), $\xi(t)$. At the end of this section, we will find $y(t)$ and $\xi(t)$ are equivalent.

As discussed above, the transfer function is the impulse response of the system. We can interpret that the impulse response of (6.48) corresponds to the homogeneous solution $\tilde{\xi}(t)$. We discussed that $\tilde{\xi}(t)$ can take the following form.

$$\tilde{\xi}(t) = a \sin \omega_0 t \tag{6.54}$$

So, we can put $w(t - \tau) = a \sin \omega_0(t - \tau)$ in (6.23). Using the source term $\alpha \sin \Omega t$ for $f(\xi)$ in (6.23), we can express the output $y(t)$ as follows.

$$
\begin{aligned}
y(t) &= a \int_0^t \alpha \sin \Omega \tau \sin \omega_0(t - \tau) d\tau \\
&= \frac{a\alpha}{2} \int_0^t \cos\{\Omega\tau - \omega_0(t - \tau)\} d\tau - \frac{a\alpha}{2} \int_0^t \cos\{\Omega\tau + \omega_0(t - \tau)\} d\tau \\
&= \frac{a\alpha}{2} \int_0^t \cos\{(\Omega + \omega_0)\tau - \omega_0 t\} d\tau - \frac{a\alpha}{2} \int_0^t \cos\{(\Omega - \omega_0)\tau + \omega_0 t\} d\tau \\
&= \frac{a\alpha}{2(\Omega + \omega_0)} \left[ \sin\{(\Omega + \omega_0)\tau - \omega_0 t\} \right]_0^t \\
&\qquad - \frac{a\alpha}{2(\Omega - \omega_0)} \left[ \sin\{(\Omega - \omega_0)\tau + \omega_0 t\} \right]_0^t \\
&= \frac{a\alpha}{2(\Omega + \omega_0)} \left[ \sin \Omega t - \sin(-\omega_0 t) \right] - \frac{a\alpha}{2(\Omega - \omega_0)} \left[ \sin \Omega t - \sin \omega_0 t \right] \\
&= \frac{-a\alpha 2\omega_0}{2(\Omega^2 - \omega_0^2)} \sin \Omega t + \frac{a\alpha 2\Omega}{2(\Omega^2 - \omega_0^2)} \sin \omega_0 t
\end{aligned}
\tag{6.55}
$$

By substituting solution (6.55) into differential equation (6.48), we find

$$a = \frac{1}{\omega_0} \tag{6.56}$$

hence,

$$y(t) = \frac{-\alpha(\Omega/\omega_0)}{\omega_0^2 - \Omega^2} \sin \omega_0 t + \frac{\alpha}{\omega_0^2 - \Omega^2} \sin \Omega t \tag{6.57}$$

Equation (6.57) is equivalent to the general solution $\xi(t)$, (6.53). Here, the coefficients of the $\sin \omega_0 t$ term are different between (6.53) and (6.57). We can understand this difference as follows. In deriving (6.53), we characterized the coefficient $a$ as determined by the initial condition of the homogeneous solution, which we considered independent of the particular solution (the second term). On the other hand, we derived (6.57) by viewing the spring-point-mass system as a linear system using the $\sin \Omega t$ term as the driving force (input). This linear system responds to the input with its natural frequency $\omega_0$. Therefore, the homogeneous and particular solutions are not independent of each other.

### 6.5.4 Cross-Correlation

As will be discussed later in this book, often it becomes necessary to compare two images by computing the cross-correlation and performing additional manipulation using the cross-correlation. The mathematical operation known as cross-correlation [2, 12] is a useful way to evaluate how a given pair of images are alike to each other. The cross-correlation of functions $f(x)$ and $g(x)$ is defined as follows.

$$f * g \equiv \int_{-\infty}^{\infty} f^*(\tau)g(x + \tau)d\tau \tag{6.58}$$

Here $f^*(x)$ is the complex conjugate of $f(x)$.

Take some time to digest the meaning of the expression (6.58). For simplicity, consider first the case when both $f(x)$ and $g(x)$ are real functions. In this case, the integrand of (6.58) becomes the simple product of $f(x)g(x + \tau)$. The left column of Fig. 6.5 illustrates two sine functions with four different phase differences $\phi$. The right column of the figure shows the product $f(\theta)g(\theta + \phi)$ as a function of phase $\theta$.

The area of the $f(\theta)g(\theta + \phi)$ graph presented on the right column for each phase difference $\phi$ corresponds to the integral on the right-hand side of (6.58). Notice that although $f(\theta)g(\theta + \phi)$ oscillates with the same amplitude of unity for all the cases of $\phi$, the center of oscillation varies. When $\phi = \pi/2$ the oscillation is centered at zero, making the area hence $f * g$ zero. When $\phi = 0$ and $\pi/4$, the oscillation center is positive, therefore the area of the graph, hence, the cross-correlation, is positive. However, the value of $f * g$ is greater for $\phi = 0$ than $\pi/4$, indicating that the former case is more highly correlated. When $\phi = \pi$ the cross-correlation is negative, with the same absolute value as the case of $\phi = 0$.

We can normalize the cross-correlation to the case when the two functions are the most similar; in the present case when the two sine functions have no relative phase difference. The normalized cross-correlation evaluated in this manner is inserted in each graph on the right column of Fig. 6.5 (e.g., 'corr = 1.000'). The plots on the left column indicate when the normalized cross-correlation is unity the two functions are completely overlapping each other. When the normalized correlation is zero the two sine functions are off by a quarter-period and when the normalized correlation is $-1$ the two functions are completely out of phase.

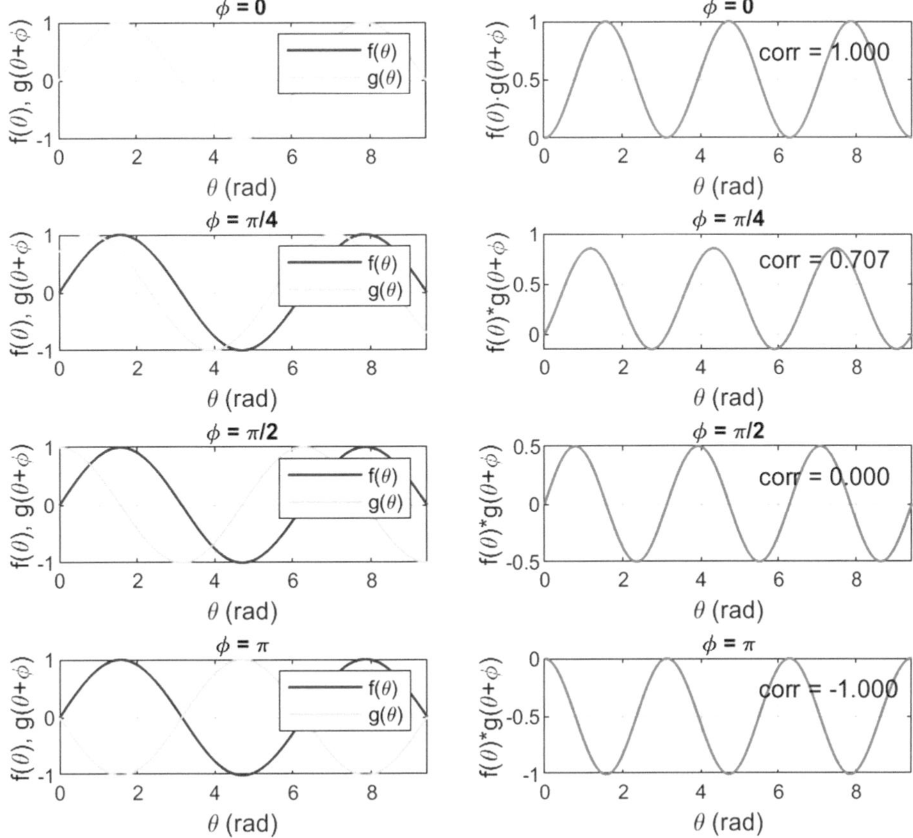

**Fig. 6.5** Cross-correlation of two real functions

Above we find that the convolution integral can be evaluated in the frequency domain more easily than in the time domain. As easily seen, there is a formulaic similarity between the convolution integral and correlation. It is anticipated that cross-correlation can also be evaluated in the frequency domain. In fact, it is much easier to evaluate cross-correlation in the frequency domain. Once it is evaluated in the frequency domain, it is easily converted back to the time or space domain by inversely Fourier transform the result. Below we derive the frequency domain expression of cross-correlation.

Consider the Fourier transform of (6.58). From the definition,

$$\mathcal{F}\{f * g\} = \int_{-\infty}^{\infty} \left[ \int_{-\infty}^{\infty} f^*(\tau)g(x+\tau)d\tau \right] e^{-j2\pi\nu x}dx \tag{6.59}$$

Putting $y = x + \tau$, hence $dx = dy$,

$$\begin{aligned}
\mathcal{F}\{f * g\} &= \int_{-\infty}^{\infty} \left[ \int_{-\infty}^{\infty} f^*(\tau)g(y)d\tau \right] e^{-j2\pi\nu(y-\tau)}dy \\
&= \int_{-\infty}^{\infty} f^*(\tau)e^{-j2\pi(-\nu)\tau}d\tau \int_{-\infty}^{\infty} g(y)e^{-j2\pi\nu y}dy \\
&= \mathcal{F}\{[f(x)]^*\}(-\nu)\mathcal{F}\{[g(x)]\}(\nu) \\
&= F^*(\nu)G(\nu)
\end{aligned} \tag{6.60}$$

Here in the last step, the above-discussed property of the Fourier transform regarding the complex conjugate is used (5.47), and the upper case represents the Fourier transform of the lower case.

## 6.6   Solving Differential Equations in Frequency Domain

Above we discussed the advantage of frequency domain analysis on linear systems. In this section, we consider solving differential equations that represent the dynamics of linear systems in the frequency domain. After discussing the representation of linear differential equations in the frequency domain briefly, we apply the concept to actual physical systems. We will find that in the frequency domain, we can view the differential equation as the transfer function that 'transfers' the effect of input to output.

### 6.6.1   Differential Equation as a Transfer Function

Analysis of dynamic systems essentially means solving the equation of motion that governs the system. In other words, we need to solve a differential equation that represents the equation of motion and find the motion of the object as a solution to the differential equation. In engineering applications, often finding a particular solution is important. Fourier transform is conveniently used to find a particular solution of a linear differential equation, i.e., find the solution for a given source term of the differential equation. In the frequency domain, the source term is considered to be the input to the system and the solution is the output.

Consider the following differential equation.

$$a\ddot{x}(t) + b\dot{x}(t) + cx(t) = y(t) \tag{6.61}$$

The procedure to solve the differential equation consists of the following steps. First, convert the given differential equation from the time domain to the frequency domain. The actual action of this step is to take the Fourier transform of each term of the differential equation. With the property of Fourier transform regarding differentiation (Sect. 5.5.7), all the terms on the left-hand side can be expressed in the form of a product to the Fourier transform of the solution $(X(\omega))$ and $j\omega \equiv s$ or $X(\omega)$ itself.

$$as^2 X(\omega) + bs X(\omega) + cX(\omega) = \left(as^2 + bs + c\right) X(\omega) = Y(\omega) \tag{6.62}$$

Here $X(\omega)$ is the Fourier transform of the solution, $Y(\omega)$ is the Fourier transform of the input (source term). The second step is to express the frequency-domain differential equation (6.62) in the following form.

$$X(\omega) = \frac{Y(\omega)}{\left(as^2 + bs + c\right)} = Y(\omega)H(\omega) \tag{6.63}$$

Here $H(\omega)$ represents the transfer function in the sense that the input $Y(\omega)$ is transferred to the solution $X(\omega)$ via the linear system represented by $H(\omega)$.

The final step is to solve this equation for $X(\omega)$, and inverse Fourier transform of it to find the solution $x(t)$ in the time domain.

$$x(t) = \mathcal{F}^{-1}\left(X(\omega)\right) \tag{6.64}$$

Note that while this method is convenient to find the temporal behavior of the solution, the initial condition is undetermined.

## 6.6.2 Frequency Domain Analysis of Forced Oscillation

As the first example of actual physical systems, consider the dynamics of the forced oscillation of a spring-point-mass system that we called Damped Driven Harmonic Oscillation in Sect. 2.4. We can start from the equation of motion (2.65).

$$\frac{d^2\xi(t)}{dt^2} + 2\beta\frac{d\xi(t)}{dt} + \omega_0^2\xi(t) = \frac{f_{ex}}{m}\sin\Omega t \tag{2.65}$$

Consider Fourier transform each term of Eq. (2.65). Remember that in Sect. 5.5.7 we discussed that in the Fourier domain, the differentiation of function $g(t)$ is the multiplication of angular frequency and the imaginary unit to the Fourier transform of the function. $i\omega G$.

$$\mathcal{F}\left\{\dot{g}(t)\right\} = i\omega G \tag{5.50}$$

Apply this rule to the differential equation (2.65) and obtain the following differential equation in the frequency domain.

$$s^2 \, \Xi(\omega) + 2\beta s \, \Xi(\omega) + \omega_0^2 \Xi(\omega) = \left(s^2 + 2\beta s + \omega_0^2\right) \Xi(\omega) = \frac{F_{ex}}{m} \tag{6.65}$$

Here $s = i\omega$, and $\Xi(\omega)$ and $F_{ex}(\omega)$ are the Fourier transforms of displacement $\xi(t)$ and $f_{ex}$, respectively.

By solving (6.65) for $\Xi(\omega)$, we find

$$\Xi(\omega) = \frac{F_{ex}}{m \left(s^2 + 2\beta s + \omega_0^2\right)} \equiv H(\omega) F_{ex} \tag{6.66}$$

Here the quantity $H(\omega)$ is the transfer function (in this particular case, the driving force to displacement transfer function), and represents the frequency dependence of the linear system that converts the driving force to the displacement of the point mass. In this context, the driving force is the input to the linear system, the displacement is the output from the linear system, and the transfer function characterizes the linear system.

$$H(\omega) = \frac{1}{m \left(s^2 + 2\beta s + \omega_0^2\right)} \tag{6.67}$$

Figure 6.6 shows the transfer function of the linear system represented by (6.67) with the mass 1 kg, natural frequency 500 Hz, and decay constant 40 and 100 1/s. The top plot shows the magnitude of the transfer function and the bottom shows its phase.

It should be noted that Fig. 6.6 is identical to Fig. 2.8 which plots the amplitude and phase of forced oscillation of the same spring-mass system as a function driving frequency. The magnitude corresponds to the amplitude $A$ in expression (2.73), and the phase corresponds to $-\delta$ in (2.74) of the harmonic response with the amplitude of unity for the input (external force). This indicates that the transfer function characterizes the linear system's harmonic response. In harmonic response, the input signal has a single frequency, and therefore the situation corresponds to the transfer function at that frequency. (In Fig. 2.8 I used "$f_{ex} = 1$ N" in the figure title to emphasize that this figure represents the mass's oscillatory behavior when driven by the external force with the amplitude of 1 N. I label the plots in Fig. 6.6 "m = 1 (kg)" to emphasize that the figure characterizes the oscillatory system consisting of 1 kg mass in the frequency domain.)

It is seen in Fig. 6.6 that (1) the amplitude is maximized near the natural frequency, (2) the amplitude peak is higher with its profile sharper when the decay constant, hence the damping coefficient, is lower, (3) the phase changes from zero toward $-\pi$ rad as the driving frequency approach the resonance frequency from the lower frequency side, and (4) this change in the phase is steeper with the lower damping coefficient. (2) explicitly indicates that a higher damping coefficient makes the energy transfer from the driving force to the oscillation of the linear system less efficient, as we discussed with Fig. 2.7 in Sect. 2.4.

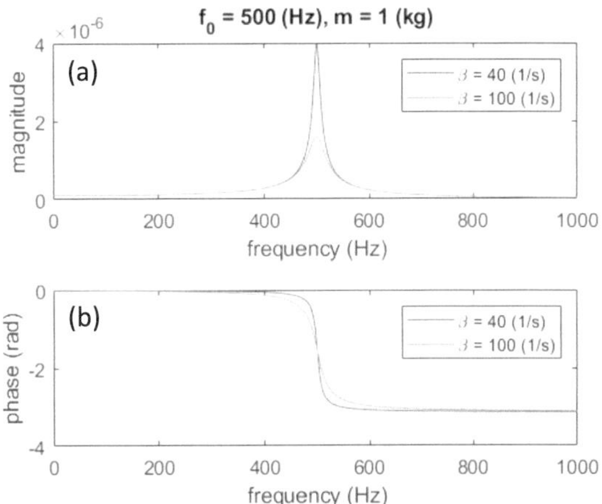


**Fig. 6.6** Transfer function of the linear system represented by mass 1 kg, natural frequency 500 Hz, and decay constant 40 1/s and 100 1/s

## Resonance

At the end of Sect. 2.4, we briefly touched on the concept of resonance. Here we consider this topic in more detail.

The near resonance behaviors observed in Fig. 6.6 can be summarized as follows.

(a) The amplitude is maximized when the linear system is driven near the natural frequency. The denominator of (2.73) takes the minimum value when the driving frequency $\Omega = \sqrt{\omega_0^2 - 2\beta^2}$. In other words, the damping mechanism lowers the resonant frequency from the natural frequency. In most oscillatory systems, however, the natural frequency is significantly higher than the damping effect ($\omega_0 >> \beta$), and hence the resonant frequency is close to the natural frequency ($\Omega \approx \omega_0$). This is why the amplitude appears to be at the maximum when the driving frequency is equal to the natural frequency in Fig. 6.6. From the viewpoint of energy, the work done by the driving force is transferred to the oscillatory linear system more efficiently as the driving frequency approaches the natural frequency. We can interpret this energy transfer as the driving (excitation) energy is absorbed by the oscillator. A certain atom absorbs light of a certain frequency can be explained with this mechanism.

(b) As the decay constant increases, the peak amplitude decreases and at the same time the bell-shaped profile broadens.
The second term in the denominator of (2.73) is the product of the decay constant and driving frequency. As the decay constant increases, this term lowers the amplitude and this tendency is greater near the amplitude peak. For a given mass $m$, an increase in the decay constant means an increase in the damping coefficient. As the damping coefficient increases, the energy transfer from the driving force to the oscillator becomes less efficient as more energy is dissipated through the damping mechanism. Since the oscillator receives less energy in the oscillatory motion at the natural frequency, the amplitude peak decreases.

(c) The phase changes from zero toward $-\pi$ rad as the driving frequency approaches the resonance frequency from the lower frequency side.

When the external force drives the linear system at a frequency lower than the natural frequency, the system follows the driving motion. That is why the phase delay is small. At the zero Hz limit, the phase delay is exactly zero. This indicates that if we apply a constant (non-oscillatory) force on the mass, it follows the applied force with no phase delay. (Consider that you grab the top of a pendulum with your hand and walk very slowly. You can easily imagine that the bob of the pendulum will follow you at the same speed.) When the driving frequency is higher than the natural frequency, the mass cannot follow the driving action, because it tends to oscillate at its natural frequency. Consequently, the mass tends to move out of phase relative to the driving motion. At the high-frequency limit, the relative motion becomes exactly out of phase. At the resonance, the phase difference is $-\pi/2$, or exactly at the halfway point of 0 and $-\pi$.

Algebraically, this can be explained with (2.74) as follows. When the driving frequency $\Omega = 0$, the numerator of (2.74) becomes zero and therefore $\delta = 0$. At the resonance, $\Omega = \omega_0$ makes the denominator of (2.74) zero making $\tan \delta \to \infty$ hence $\delta = \pi/2$. As the driving frequency increases from $\omega_0$, the denominator of (2.74) decreases approaching zero from $\delta = -\pi/2$, making $\delta \to -\pi$ at the $\Omega \to \infty$ limit.

(d) As the decay constant increases, the transition of 0 to $-\pi$ phase difference becomes less steep.

This indicates that with the increase in the damping effect, the oscillator's resonant effect becomes less prominent.

### 6.6.3  Transfer Function of an LC Circuit

As another example of a transfer function representing physical systems, consider an electric circuit consisting of a capacitor $C$ and inductor $L$. It should be noted that by interpreting what quantities as the input and output, we can define different transfer functions for the same physical systems.

First, consider in Fig. 6.7 the transfer function between the current flowing through the series connection of the capacitor and inductor and the voltage across the ends of the series

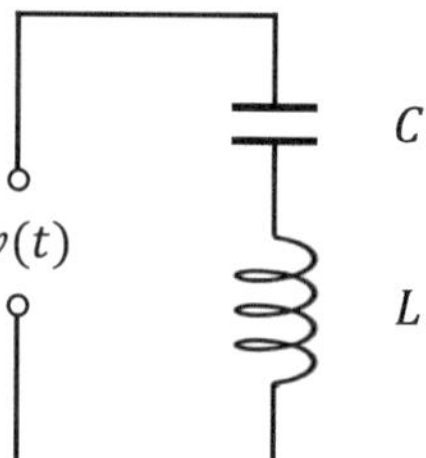


**Fig. 6.7** Series connection of a capacitor and inductor. The transfer function represents the ratio of current in the secondary coil to voltage input

connection, $v(t)$. Considering the voltage drops across the capacitor and inductor, we can express $v(t)$ as follows.

$$v(t) = v_c(t) + v_L(t) = \frac{q(t)}{C} + L\frac{di(t)}{dt} = \frac{1}{C}\int i(t)dt + L\frac{di(t)}{dt} \tag{6.68}$$

Using the same idea as (6.62), we can rewrite (6.68) in the frequency domain as follows.

$$V(\omega) = \left(\frac{1}{sC} + sL\right) I(\omega) \tag{6.69}$$

Here $V(\omega)$ and $I(\omega)$ are the Fourier transform of the voltage and current. Viewing the voltage to be the input to the system and the current as the output, we can put (6.69) in the same form as (6.63).

$$I(\omega) = H(\omega)V(\omega) \tag{6.70}$$

$$H(\omega) = \frac{1}{\frac{1}{sC} + sL} = \frac{sC}{1 + s^2LC} = \frac{j\omega C}{1 - \omega^2 LC} \tag{6.71}$$

Here $s$ is replaced with $j\omega$ to make the physical meaning of the transfer function more intuitive. Equations (6.70) and (6.71) indicate that when the input voltage is DC ($\omega = 0$), the current is zero and consequently the output voltage (across the ends of the circuit) is zero. On the other hand, if the frequency of the input voltage is at the resonance ($\omega = 1/\sqrt{LC}$), the current is infinite. In reality, the internal resistance of the circuit makes the current finite at resonance. However, the output voltage is certainly maximized when the input voltage is at the resonant frequency.

   If we interpret the current flowing through the circuit to be the input and voltage at the ends be the output, the transfer function $H$ to be the reciprocal of $H$ in (6.70).

$$H(\omega) = \frac{1 + s^2LC}{sC} = \frac{1 - \omega^2 LC}{j\omega C} \tag{6.72}$$

This time DC input yields infinite output and input at the resonant frequency yields null output.

   Next, consider the transfer function between the voltage across the end of the circuit and the charge stored in the capacitor $q(t)$. Rewrite the time domian circuit equation (6.68) using $v(t)$ and $q(t)$.

$$v(t) = v_c(t) + v_L(t) = \frac{q(t)}{C} + L\frac{d^2q(t)}{dt^2} \tag{6.73}$$

The corresponding equation in the frequency domain becomes

$$V(\omega) = \left(\frac{1}{C} + s^2L\right) Q(\omega) \tag{6.74}$$

Viewing $q(t)$ as the output and $v(t)$ as the input, we obtain the following equations.

$$Q(\omega) = H(\omega)V(\omega) \tag{6.75}$$

$$H(\omega) = \frac{1}{\frac{1}{C} + s^2 L} = \frac{C}{1 + s^2 LC} = \frac{C}{1 - \omega^2 LC} \tag{6.76}$$

This time a DC input provides a constant output whereas the output is maximized when the input frequency is $\omega = 1/\sqrt{LC}$.

**Zeros and poles**

Equations (6.70)–(6.75) indicate that the transfer function of a linear system can be generally expressed in the following form.

$$H(s) = K\frac{(s - z_1)(s - z_2)}{(s - p_1)(s - p_2)} \tag{6.77}$$

If multiple linear systems are connected in series, we can multiply transfer functions of this form. It is easily seen that the resultant overall transfer function has the same general form. Here $z_i$ and $p_i$ are called the zeros and poles [13].

As an example, add a resistor to the LC circuit shown in Fig. 6.7. This modifies (6.68) and (6.69) as follows.

$$v(t) = v_c(t) + v_L(t)v_R(t) = \frac{q(t)}{C} + L\frac{di(t)}{dt} + Ri(t)$$

$$= \frac{1}{C}\int i(t)dt + L\frac{di(t)}{dt} + Ri(t) \tag{6.78}$$

$$V(\omega) = \left(\frac{1}{sC} + R + sL\right)I(\omega) \tag{6.79}$$

The corresponding transfer function from $V(\omega)$ to $I(\omega)$ takes the form of (6.77).

$$H(s) = \frac{I(\omega)}{V(\omega)} = \frac{s}{Ls^2 + Rs + \frac{1}{C}}$$

$$= \frac{s}{(s - \frac{-R+\sqrt{R^2-4L/C}}{2L})(s - \frac{-R-\sqrt{R^2-4L/C}}{2L})} \tag{6.80}$$

We see that this particular system as a zero at $s = 0$ and poles at $s = (-R + \sqrt{R^2 - 4L/C})/(2L)$ and $s = (-R - \sqrt{R^2 - 4L/C})/(2L)$.

---

## References


1. Antsaklis PJ, Michel AN (1997) Linear systems. Birkhäuser, Boston, Berlin
2. Mavko G, Mukerji T, Dvorkin J (2020) The rock physics handbook, 3rd edn. Cambridge University Press, New York. Chap. 1
3. Schiff JL (1999) The Laplace transform: theory and applications. Springer, New York

4. Dyke P (2001) An introduction to Laplace transforms and Fourier series, 2nd edn. Springer, New York

5. Nixon FE (1965) Handbook of Laplace transformation: fundamentals, applications, tables, and examples, 2nd edn. Prentice-Hall, Englewood Cliffs, NJ

6. Hirschman II, Widder DV (2005) The convolution transform. Dover, New York

7. Behl A, Puri A (2014) Convolution and applications of convolution. Int J Innov Res Technol 1(6):2122–2126

8. Basso CP (2016) Linear circuit transfer functions: an introduction to fast analytical techniques. Wiley, Hoboken, NJ

9. Åström KJ, Murray, RM (2021) Feedback systems, 2nd edn. Princeton University Press, Princeton, NJ

10. Evans LC (2010) Partial differential equations. Graduate studies in mathematics, vol 19, 2nd edn. American Mathematical Society, Rhode Island

11. Prasad BKR (2020) Structural dynamics in earthquake and blast resistant design, (Duhamel's integral). CRC Press, Boca Raton, Florida. Chap. 5

12. Bendat JS (1993) Engineering applications of correlation and spectral analysis, 2nd edn. John Wiley & Sons, New York

13. Scherbaum F (2001) Of poles and zeros, 2nd edn. Kluwer Academic Publishers, Dordrecht, Netherlands

# Kirchhoff Integral

Our goal here is to derive a wave equation in the integral form as (4.17). For simplicity, let's derive a wave equation that a wave solution of the following form satisfies.

$$\xi = \psi(x, y, z)e^{-i(\omega t - kr)} = u(x, y, z)e^{-i\omega t} \tag{A.1}$$

Assume that solution $\xi$ satisfies the wave equation in the following differential form.

$$\frac{\partial \xi^2}{\partial t^2} = v^2 \nabla^2 \xi \tag{A.2}$$

Here $v$ is the wave velocity.

Substitution of solution (A.1) into (A.2) yields the following equations.

$$\frac{\partial^2 \xi}{\partial t^2} = -\omega^2 u e^{-i\omega t} \tag{A.3}$$

$$\nabla^2 \xi = \nabla^2 u e^{-i\omega t} \tag{A.4}$$

From (A.2)–(A.4), we obtain the following differential equation for $u$.

$$\nabla^2 u + \frac{\omega^2}{v^2} u = 0 \tag{A.5}$$

At this point, we make use of Green's theorem to relate a volume integral to the surface integral over the volume. This makes sense because we are trying to derive a wave equation that describes a wave emitted from a source located inside a volume and observed at the observation point P on the surface that encloses the source. According to Green's theorem, we can relate the volume and surface integral for an arbitrary vector $\mathbf{A}$ as follows.

$$\int_\Omega \nabla \cdot \mathbf{A}\, d\Omega = \int_S \mathbf{A} \cdot \mathbf{n}\, dS \tag{A.6}$$

© The Editor(s) (if applicable) and The Author(s), under exclusive license to Springer Nature Switzerland AG 2025
S. Yoshida, *Physics and Mathematics Behind Wave Dynamics*, Synthesis Lectures on Wave Phenomena in the Physical Sciences,
https://doi.org/10.1007/978-3-031-60354-9

Here, $\Omega$ is the volume, $S$ is the surface of the volume, and $\mathbf{n}$ is the unit vector normal to the surface. Note that $\mathbf{n}$ is positive when the normal vector is outward on the surface $S$.

For the reason clarified shortly, we put $\mathbf{A} = u\nabla w$ using a spatial function $w$. With this form of $\mathbf{A}$, (A.6) becomes as follows.

$$\int_{\Omega} \nabla \cdot (u\nabla w)d\Omega = \int_{S} u\nabla w \cdot \mathbf{n}dS \tag{A.7}$$

Here, the left-hand and right-hand sides of (A.7) can be put as follows.

$$\int_{\Omega} \nabla \cdot (u\nabla w)d\Omega = \int_{\Omega} (u\nabla^2 w + \nabla u \cdot \nabla w)d\Omega$$
$$\int_{S} u\nabla w \cdot \mathbf{n}dS = \int_{S} u\frac{\partial w}{\partial n}dS$$

Therefore, we can rewrite (A.7) as follows.

$$\int_{\Omega} (u\nabla^2 w + \nabla u \cdot \nabla w)d\Omega = \int_{S} u\frac{\partial w}{\partial n}dS \tag{A.8}$$

Since Greens' theorem holds for any function whose spatial derivative is continuous, we can switch $u$ and $v$ in (A.8) and obtain the following equation.

$$\int_{\Omega} (w\nabla^2 u + \nabla w \cdot \nabla u)d\Omega = \int_{S} w\frac{\partial u}{\partial n}dS \tag{A.9}$$

Subtraction of (A.9) from (A.8) yields the following equation.

$$\int_{\Omega} (u\nabla^2 w - w\nabla^2 u)d\Omega = \int_{S} \left( u\frac{\partial w}{\partial n} - w\frac{\partial u}{\partial n} \right) dS \tag{A.10}$$

Now consider that $w$ satisfies the same wave equation as $u$.

$$\nabla^2 w + \frac{\omega^2}{v^2}w = 0 \tag{A.11}$$

Then from (A.5) and (A.11) we find

$$u\nabla^2 w = w\nabla^2 u$$

and therefore the left-hand side of (A.10) is equal to zero. Thus,

$$\int_{S} \left( u\frac{\partial w}{\partial n} - w\frac{\partial u}{\partial n} \right) dS = 0 \tag{A.12}$$

We select the following function for $w$.

$$w(x, y, z) = \frac{e^{ikr}}{r} \tag{A.13}$$

**Fig. A.1** Closed volume to apply Green's theorem without divergence to infinity

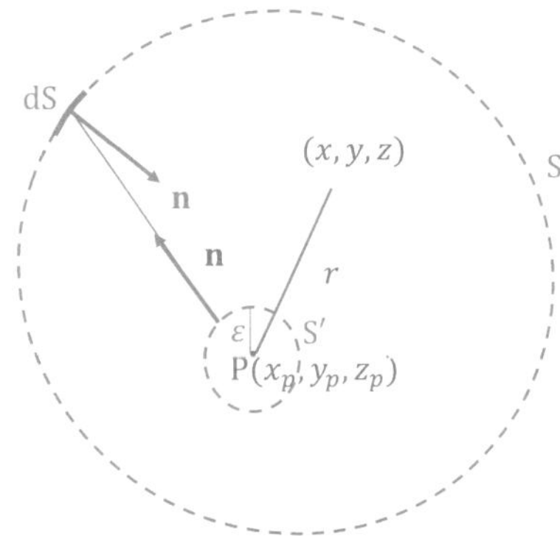


Here $k$ and $r$ are the propagation constant and radial distance from the observation point P.

$$k = \frac{\omega}{v}$$

$$r = \sqrt{(x - x_p)^2 + (y - y_p)^2 + (z - z_p)^2}$$

It is easily proved that (A.13) satisfies the wave equation (A.11). However, when we consider a closed volume in the application of Green's theorem we cannot include point P because at $r = 0$, $w$ diverges to infinity. To avoid this problem we consider a sphere with an infinitesimally small radius $\epsilon$ around point P and set up a closed volume between the surface of this small sphere and the outer surface as illustrated in Fig. A.1.

With this setting the surface integration is taken over the inner surface $S'$ and the outer surface S. (A.12) becomes as follows.

$$\int_S \left( u \frac{\partial w}{\partial n} - w \frac{\partial u}{\partial n} \right) dS + \int_{S'} \left( u \frac{\partial w}{\partial n} - w \frac{\partial u}{\partial n} \right) dS'$$
$$= \int_S \left( u \frac{\partial}{\partial n} \frac{e^{ikr}}{r} - \frac{e^{ikr}}{r} \frac{\partial u}{\partial n} \right) dS + \int_{S'} \left( u \frac{\partial}{\partial n} \frac{e^{ikr}}{r} - \frac{e^{ikr}}{r} \frac{\partial u}{\partial n} \right) dS' = 0 \qquad \text{(A.14)}$$

In going to the second line of (A.14), (A.13) is used. For simplicity, introduce normal vector $\mathbf{n}(l, m, n)$ inward on the two surfaces. (Normally normal vectors are outward in application of Green's theorem. We used an outward normal vector in the derivation of (A.7). However, changing the direction of the normal vector from outward positive to inward positive changes only the sign of the entire (A.14). Since the value of this equation is zero, the change in the sign does not change the equation).

The derivative $\partial/\partial n$ that appears in (A.14) is the partial differentiation with respect to the unit length along the normal direction to the surface. It is useful to express this partial derivative in the following form.

$$\frac{\partial}{\partial n} = \frac{\partial x}{\partial n} \frac{\partial}{\partial x} + \frac{\partial y}{\partial n} \frac{\partial}{\partial y} + \frac{\partial z}{\partial n} \frac{\partial}{\partial z} = \alpha \frac{\partial}{\partial x} + \beta \frac{\partial}{\partial y} + \gamma \frac{\partial}{\partial z} \qquad \text{(A.15)}$$

**Fig. A.2** Unit vector and its components

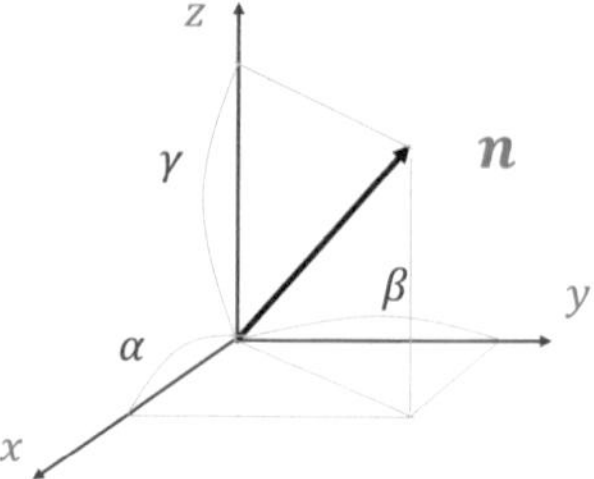


In (A.15), $\partial x_i/\partial n$ $(x_i = x, y, z)$ are the $x_i$ component of the unit normal vector $\mathbf{n}$ (Fig. A.2). In the rightmost term of (A.15), $\alpha$, $\beta$, $\gamma$ represent these components. Noting that $\mathbf{n}$ is inward with respect to the enclosed volume $\Omega$, we can express the radial vector $\mathbf{r}$ as follows.

$$\mathbf{r} = (x - x_p)\hat{i} + (y - y_p)\hat{j} + (z - z_p)\hat{k} \tag{A.16}$$

Hence,

$$r = |\mathbf{r}| = \sqrt{(x - x_p)^2 + (y - y_p)^2 + (z - z_p)^2} \tag{A.17}$$

It is convenient to express the following derivatives using (A.17).

$$\frac{\partial r}{\partial x} = \frac{\partial \sqrt{(x - x_p)^2 + (y - y_p)^2 + (z - z_p)^2}}{\partial x}$$
$$= \frac{2(x - x_p)}{2\sqrt{(x - x_p)^2 + (y - y_p)^2 + (z - z_p)^2}} = \frac{r_x}{r} \tag{A.18}$$

$$\frac{\partial r}{\partial y} = \frac{\partial \sqrt{(x - x_p)^2 + (y - y_p)^2 + (z - z_p)^2}}{\partial y}$$
$$= \frac{2(y - y_p)}{2\sqrt{(x - x_p)^2 + (y - y_p)^2 + (z - z_p)^2}} = \frac{r_y}{r} \tag{A.19}$$

$$\frac{\partial r}{\partial z} = \frac{\partial \sqrt{(x - x_p)^2 + (y - y_p)^2 + (z - z_p)^2}}{\partial z}$$
$$= \frac{2(z - z_p)}{2\sqrt{(x - x_p)^2 + (y - y_p)^2 + (z - z_p)^2}} = \frac{r_z}{r} \tag{A.20}$$

Here $r_x$, $r_y$ and $r_z$ are the $x$, $y$ and $z$ components of vector $\mathbf{r}$ (A.16). Using (A.18)–(A.20) and (A.15), we find as follows.

$$\frac{\partial r}{\partial n} = \alpha\frac{\partial r}{\partial x} + \beta\frac{\partial r}{\partial y} + \gamma\frac{\partial r}{\partial z} = \frac{\alpha r_x}{r} + \frac{\beta r_y}{r} + \frac{\gamma r_z}{r} = \frac{\mathbf{n} \cdot \mathbf{r}}{r} \tag{A.21}$$

So,

$$\frac{\partial}{\partial n}\left(\frac{1}{r}\right) = \frac{\partial}{\partial r}\left(\frac{1}{r}\right)\frac{\partial r}{\partial n} = -\frac{1}{r^2}\frac{\mathbf{n}\cdot\mathbf{r}}{r} \tag{A.22}$$

Using (A.21),

$$\frac{\partial}{\partial n}\left(\frac{e^{ikr}}{r}\right) = \frac{\partial}{\partial r}\left(\frac{e^{ikr}}{r}\right)\frac{\partial r}{\partial n} = \frac{e^{ikr}}{r}\left(ik - \frac{1}{r}\right)\frac{\partial r}{\partial n}$$

$$= \frac{e^{ikr}}{r^2}\left(ik - \frac{1}{r}\right)\mathbf{n}\cdot\mathbf{r} \tag{A.23}$$

On the surface $S'$ we can express as follows.

$$\mathbf{n}\cdot\mathbf{r} = \epsilon \tag{A.24}$$

$$dS' = \epsilon^2 d\phi d\theta \sin\phi \tag{A.25}$$

Here $\phi$ and $\theta$ are the polar angles and azimuthal angle of the polar coordinate system associated.

With these expressions, the surface integration for $S'$ in the $\epsilon \to 0$ limit becomes as follows.

$$\int_{S'}\left(u\frac{\partial}{\partial n}\frac{e^{ikr}}{r} - \frac{e^{ikr}}{r}\frac{\partial u}{\partial n}\right)dS'$$

$$= \lim_{\epsilon\to 0}\iint_{S'}\left\{u\frac{e^{ik\epsilon}}{\epsilon}\left(ik - \frac{1}{\epsilon}\right) - \frac{e^{ik\epsilon}}{\epsilon}\frac{\partial u}{\partial n}\right\}dS'$$

$$= -\int_{-\pi}^{\pi}\int_{0}^{2\pi} u(x_p, y_P, z_P)d\theta \sin\phi d\phi$$

$$= -4\pi u(x_p, y_p, z_p) \tag{A.26}$$

From (A.14) and (A.26),

$$u(x_p, y_p, z_p) = \frac{1}{4\pi}\int_{S}\left\{u\frac{\partial}{\partial n}\left(\frac{e^{ikr}}{r}\right) - \frac{e^{ikr}}{r}\frac{\partial u}{\partial n}\right\}dS \tag{A.27}$$

From (A.1),

$$\xi(x_p, y_p, z_p, t) = u(x_p, y_p, z_p)e^{-i\omega t}$$

$$= \frac{e^{-i\omega t}}{4\pi}\int_{S}\left\{u\frac{\partial}{\partial n}\left(\frac{e^{ikr}}{r}\right) - \frac{e^{ikr}}{r}\frac{\partial u}{\partial n}\right\}dS$$

$$= \frac{e^{-i\omega t}}{4\pi}\int_{S}\left\{ue^{ikr}\frac{\partial}{\partial n}\left(\frac{1}{r}\right) + \frac{ik}{r}ue^{ikr}\left(\frac{\partial r}{\partial n}\right) - \frac{e^{ikr}}{r}\frac{\partial u}{\partial n}\right\}dS$$

$$= \frac{1}{4\pi}\int_{S}\left\{ue^{-i(\omega t - kr)}\frac{\partial}{\partial n}\left(\frac{1}{r}\right) + \frac{ik}{r}ue^{-i(\omega t - kr)}\left(\frac{\partial r}{\partial n}\right) - \frac{e^{-i(\omega t - kr)}}{r}\frac{\partial u}{\partial n}\right\}dS \tag{A.28}$$

$$\frac{\partial \xi}{\partial t} = -i\omega u e^{-i\omega t} \tag{A.29}$$

$$\frac{\partial \xi}{\partial n} = \frac{\partial u}{\partial n} e^{-i\omega t} \tag{A.30}$$

Using $k = \omega/v$ from (A.29) and (A.30) we find the derivatives at $t = t - r/v$ as follows.

$$\left(\frac{\partial \xi}{\partial t}\right)_{t-r/v} = -ikvue^{-i\omega(t-r/v)} = -i\omega u e^{-i(\omega t - kr)} \tag{A.31}$$

$$\left(\frac{\partial \xi}{\partial n}\right)_{t-r/v} = \frac{\partial u}{\partial n} e^{-i\omega(t-r/v)} = \frac{\partial u}{\partial n} e^{-i(\omega t - kr)} \tag{A.32}$$

Substituting (A.31) and (A.32) into (A.28), we obtain the following expression.

$$\xi(x_p, y_p, z_p, t) = u(x_p, y_p, z_p)e^{-i\omega t}$$

$$= \frac{1}{4\pi} \int_S \left\{ \xi \frac{\partial}{\partial n}\left(\frac{1}{r}\right) - \frac{1}{rv}\left(\frac{\partial r}{\partial n}\right)\left(\frac{\partial \xi}{\partial t}\right) - \frac{1}{r}\left(\frac{\partial \xi}{\partial n}\right) \right\}_{t-(r/v)} dS \tag{A.33}$$

Expression (A.33) is the same as (4.17).

# Fourier Transform of Some Functions B

So far we have discussed the Fourier transform abstractly. In this section, we discuss the Fourier transform of some functions that we often use in various engineering or physical applications.

## B.1 Delta Function $\delta(t)$

The delta function [1], or the unit impulse function, $\delta(t)$ is defined as follows.

$$\delta(t) = \begin{cases} 0 & t \neq 0 \\ \infty & t = 0 \end{cases} \tag{B.1}$$

$$\int_{-\infty}^{\infty} \delta(t)dt = 1 \tag{B.2}$$

Equation (B.1) defines the value of the function and (B.2) specifies that the area of the function is unity. With the finite area and infinite value at $t = 0$, it follows that the function has a null width along the time axis. Figure B.1a shows a sample $\delta(t)$. Note that the vertical axis represents the area of the function.

The Delta function is convenient to extract a function $f(t)$ at a given time $t$. Consider first the following integration.

$$\int_{-\infty}^{\infty} \delta(t)f(t)dt = f(0) \int_{-\infty}^{\infty} \delta(t)dt = f(0) \tag{B.3}$$

Since $\delta(t) = 0$ at any $t \neq 0$, the integrand of (B.3) is zero at all values of $t$ except $t = 0$. Therefore, in the integrand the function $f(t)$ is meaningful only at $t = 0$. Consequently, we

© The Editor(s) (if applicable) and The Author(s), under exclusive license to Springer Nature Switzerland AG 2025
S. Yoshida, *Physics and Mathematics Behind Wave Dynamics*, Synthesis Lectures on Wave Phenomena in the Physical Sciences,
https://doi.org/10.1007/978-3-031-60354-9

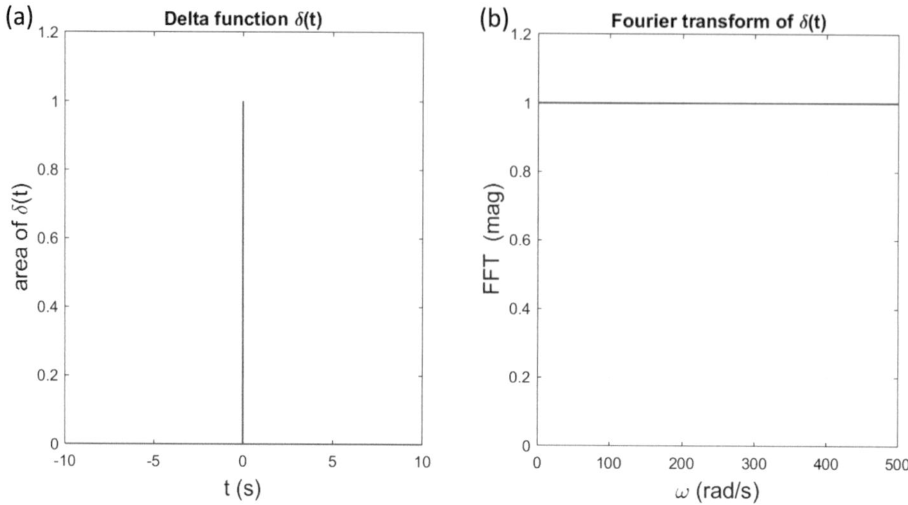

**Fig. B.1** Delta function on **a** time axis and **b** frequency axis

can pull out $f(0)$ from the integration as indicated by the second part of (B.3). Thus, we can use (B.2) to find the value of this integration to be equal to $f(0)$.

By shifting $\delta(t)$ on the time axis by $t_0$ as $\delta(t - t_0)$, we can create a situation where the product with a given function has a value only at $t = t_0$ (because the value of the Delta function is zero except $t = t_0$). Thus, repeating the same argument as (B.3), we can find the following equality.

$$\int_{-\infty}^{\infty} \delta(t - t_0) f(t) dt = f(t_0) \int_{-\infty}^{\infty} \delta(t) dt = f(t_0) \tag{B.4}$$

This property is useful in signal processing. (See Sect. 6.3 as an example.)

Evaluate the Fourier Transform of $\delta(t)$ according to (5.31).

$$\mathcal{F}\{\delta(t)\} = \int_{0}^{\infty} \delta(t) e^{-j\omega t} dt = e^{-j\omega t}|_{t=0} = 1 \tag{B.5}$$

Going through the second equal sign in (B.5), the property of the Delta function (B.3) is used.

Equation (B.5) indicates that an impulse function contains all the frequency components equally. As an interesting example, consider the classical technique to find a sweet watermelon. We tap the fruit and hear the resulting sound. By the tapping action, a stress wave containing all the frequency components propagates through the fruit. Some frequency component(s) represent sugar in the meat of the fruit. From experience, we know the corresponding sound. We do not want to miss this important frequency component(s) when examining a watermelon. Therefore, it is important to assure that the tapping action generates a Delta function.

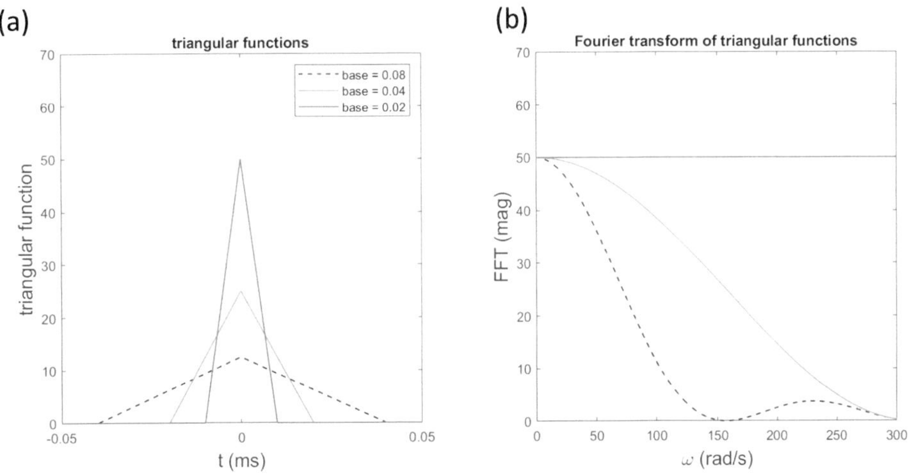



**Fig. B.2** Triangular function on **a** time axis and **b** frequency axis

Note: A type of noise that contains all frequency components is known as white noise [2]. However, an impulse signal is deterministic and not considered as white noise[1]

There are other ways to express the Delta function. These are useful when we need to numerically generate a Delta function. Here we discuss some of those functions.

### (1) Triangular Function

Figure B.2a shows triangular shapes mimicking the delta function with various base widths. The time step (the time resolution) is 0.01 s. The area of the triangle is kept at unity for all the cases. Here the narrowest case is when the base of the triangle is set at the minimum (at the time step). Figure B.2b is the corresponding Fourier transforms (magnitude) obtained by means of the Fast Fourier Transform (FFT) algorithm with the frequency resolution of 0.05 Hz, or $\pi/10$ rad/s. Note that when the base width is at the minimum, the Fourier transform is flat (constant) on the frequency axis, approaching the theoretical value indicated by (B.5) and Fig. B.1b. As the base width increases, the higher frequency components decrease. Figure B.3 plots the same triangular function as Fig. B.2 with the 10 times finer time step (0.001 s) and the same frequency resolution (0.05 Hz). Although the shapes of the time domain and frequency domain signals are the same as Fig. B.2, the absolute value of the signals are different. In the time domain, the peak value at the apex of the triangle is 10 times higher than Fig. B.2. This is because the base of the triangle in Fig. B.3 is 10 times smaller than Fig. B.2. Since both triangles have the same area (of unity), the height of the triangle with the 10 times smaller base is 10 times greater.

---

[1] See https://ccrma.stanford.edu/~jos/sasp/Why_Impulse_Not_White.html.

(a)

(b)

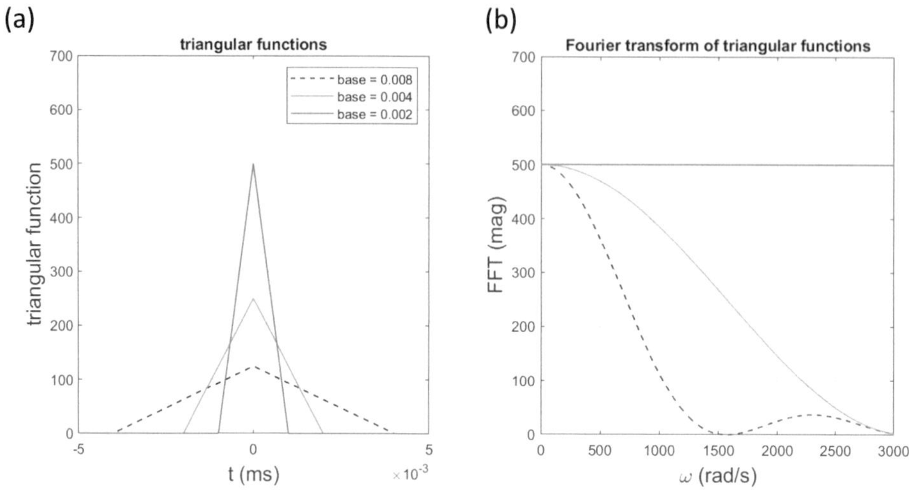


**Fig. B.3** Triangular function on **a** time axis and **b** frequency axis

The difference in the frequency domain signals comes from the difference in the highest harmonics considered in the two cases. Being inversely proportional to the time step, the maximum frequency of the signal in Fig. B.3 ($1/0.001 = 1$ kHz) is 10 times higher than Fig. B.2 ($1/0.01 = 100$ Hz.) Therefore, in terms of the proportion to the maximum frequency, the frequency value in Fig. B.2 is equivalent to the 10 times greater value in Fig. B.3. For instance, 25 Hz (157 rad/s) is 25% of the highest frequency in Fig. B.2 whereas 250 Hz (1570 rad/s) is 25% of the highest frequency in Fig. B.3.

This argument about the frequency size provides us with some insight regarding the shape of the Fourier spectrum of triangle functions. The Fourier spectrum of the triangle with the widest base in Fig. B.2 drops to zero approximately at 150 rad/s (called the drop-off frequency.) The corresponding frequency in Fig. B.3 is 1500 rad/s. This indicates that the shape of the spectrum does not represent an aspect of physics quantitatively because spectrum drops off when the frequency is a certain value relative to the maximum frequency, not at a specific frequency in Hz. The absolute frequency does not depend on the time step (as long as it is short enough to represents the function properly.) The shape of the spectrum that represents any physical phenomena is represented by the absolute value of the frequency. (See Sect. 6.6 for simple examples).

**(2) Sinc Function**

The delta function can be approximated by the limit case of the following function.

$$\delta(t) = \lim_{\epsilon \to 0} \frac{1}{\pi} \frac{\sin(t/\epsilon)}{t} \tag{B.6}$$

Here the function $\sin(t)/t$ is known as the normalized Sinc function [3], $sinc(t)$.

$$sinc(t) = \frac{1}{\pi}\frac{\sin(t/\epsilon)}{t} \tag{B.7}$$

The factor $1/\pi$ is the normalization factor that makes the integral of $sinc(t)$ for $-\infty < t < \infty$ be unity. We can prove that the integration of the right-hand side of (B.6) for $-\infty < t < \infty$ is unity regardless of the value of $\epsilon$. (See the end of this section for this integration.)

$$\int_{-\infty}^{\infty} \lim_{\epsilon \to 0} \frac{1}{\pi}\frac{\sin(t/\epsilon)}{t} = 1 \tag{B.8}$$

Figure B.4a shows the sinc function in the form of (B.6) for various $\epsilon$ on the time axis. As seen clearly, the width of the lobe located around $t = 0$ becomes sharper as the parameter $\epsilon$ approaches zero. This along with the integral being unity, (B.8), justifies the approximation of the delta function by expression (B.6).

Figure B.4b shows the magnitude of the FFT of the three sinc functions shown in Fig. B.4a. Notice that as $\epsilon$ decreases the FFT magnitude becomes broader, approaching the theoretical shape of a step-function.

**Integral of Sinc Function**

Define the following function $f(t)$.

$$f(t) \equiv \int_{-\infty}^{\infty} \frac{\sin(tx)}{x} dx \tag{B.9}$$

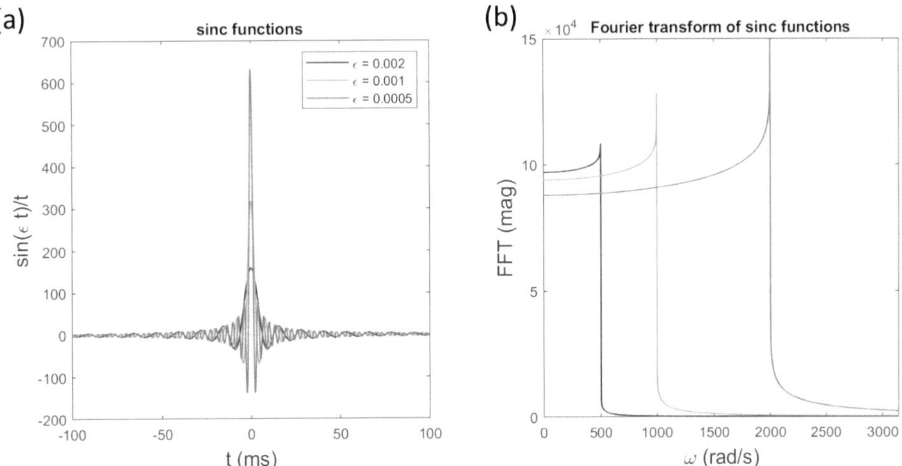


**Fig. B.4** Sinc function on **a** time axis and **b** frequency axis

Consider the Laplace transform of $f(t)$.

$$\mathcal{L}\{f(t)\} = \int_0^\infty f(t)e^{-st}dt = \int_0^\infty \int_{-\infty}^\infty \frac{\sin(tx)}{x} dx\, e^{-st}dt$$

$$= \int_{-\infty}^\infty \frac{1}{x} \left\{ \int_0^\infty \sin(tx)e^{-st}dt \right\} dx$$

$$= \int_{-\infty}^\infty \frac{1}{x} \frac{x}{s^2 + x^2} dx$$

$$= \int_{-\infty}^\infty \frac{1}{s^2 + x^2} dx = \oint_C \frac{1}{z^2 + s^2} dz$$

$$= \oint_C \frac{1}{z - is} \frac{1}{z + is} dz \tag{B.10}$$

From Cauchy's integral formula,

$$g(a) = \frac{1}{2\pi i} \oint_C \frac{g(z)}{z - a} dz \tag{B.11}$$

Putting

$$g(z) = \frac{1}{z + is} \tag{B.12}$$

$$a = is \tag{B.13}$$

into (B.11), we find

$$g(is) = \frac{1}{is + is} = \frac{1}{2is} = \frac{1}{2\pi i} \oint_C \frac{1}{z - is} \frac{1}{z + is} \tag{B.14}$$

and, hence, referring to (B.10) we find as follows.

$$\mathcal{L}\{f(t)\} = \oint_C \frac{1}{z - is} \frac{1}{z + is} = (2\pi i)\frac{1}{2is} = \frac{\pi}{s} \tag{B.15}$$

On the other hand, the Laplace transform of $\pi$ is

$$\mathcal{L}\{\pi\} = \pi \int_0^\infty e^{-st}dt = \pi \left[ \frac{1}{-s}e^{-st} \right]_0^\infty = \frac{\pi}{s} \tag{B.16}$$

Comparing (B.15) and (B.16) we find

$$f(t) = \int_{-\infty}^\infty \frac{\sin(tx)}{x} dx = \pi \tag{B.17}$$

Note that (B.17) is true for any $t > 0$ including $t = 1/\epsilon$. This leads to (B.8).

### (3) Lorentzian

The delta function can be expressed as a limit case of the Lorentz function (Lorentzian or Cauchy distribution [4, 5]) as well.

$$\delta(t) = \lim_{\epsilon \to 0} \frac{1}{\pi} \frac{\epsilon}{t^2 + \epsilon^2} \tag{B.18}$$

The Lorentzian represents a number of natural phenomena such as the power absorption spectrum of an RLC circuit (an electric circuit consisting of a resistor R, capacitor C, and inductor L), the spectral intensity of spontaneous emission from an atomic system [6], and therefore an important distribution function in natural science. It is worthwhile taking a moment here to discuss (B.18).

Consider integrating $\epsilon/(t^2 + \epsilon^2)$ appearing on the right-hand side of (B.18). Rewrite the fraction as follows.

$$\int_{-\infty}^{\infty} \frac{\epsilon}{t^2 + \epsilon^2} dt = \int_{-\infty}^{\infty} \frac{\epsilon}{\epsilon^2(1 + (t/\epsilon)^2)} dt \tag{B.19}$$

Substituting $t/\epsilon$ with $\tan \theta$, noting

$$1 + \tan^2 \theta = \frac{1}{\cos^2 \theta}$$

$$dt = \frac{\epsilon}{\cos^2 \theta} d\theta$$

we can perform the integration as follows.

$$\int_{-\infty}^{\infty} \frac{\epsilon}{t^2 + \epsilon^2} dt = \int_{-\infty}^{\infty} \frac{\epsilon}{\epsilon^2(1 + \tan^2 \theta)} dt = \int_{-\pi/2}^{\pi/2} d\theta = \pi \tag{B.20}$$

From this we find the normalized Lorentzian $L_n(t)$ takes the following form.

$$L_n(t) = \frac{1}{\pi} \frac{\epsilon}{t^2 + \epsilon^2} \tag{B.21}$$

From (B.18) and (B.21), we find that the delta function is the limit case of the normalized Lorentzian.

$$\delta(t) = \lim_{\epsilon \to 0} L_n(t) = \lim_{\epsilon \to 0} \frac{1}{\pi} \frac{\epsilon}{t^2 + \epsilon^2} \tag{B.22}$$

Figure B.5a, b respectively plot Lorentzian in the time and frequency domains with several $\epsilon$. Like the other cases discussed above, the narrower the function in the space domain the Fourier transform magnitude evaluated by the FFT algorithm becomes flatter on the frequency axis approaching the exact spectrum (B.5).

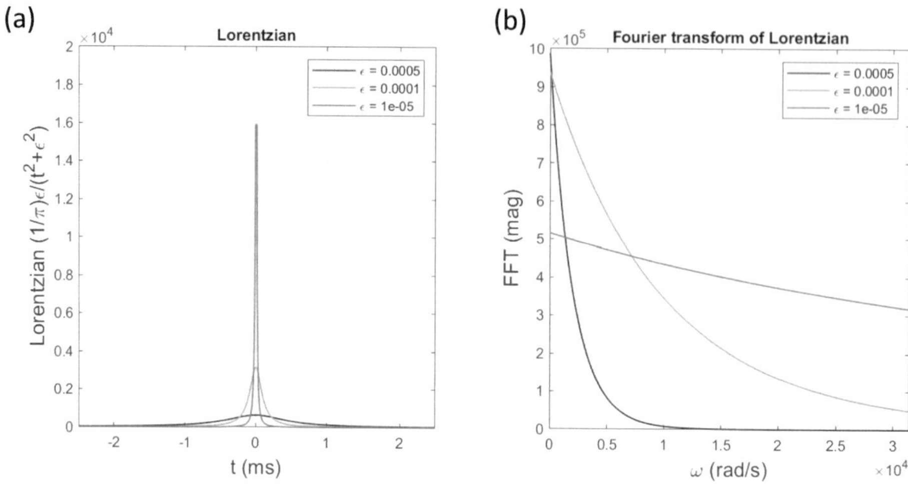

**Fig. B.5** Loretzians on **a** time axis and **b** frequency axis

## B.2    Unit Step Function $E(t)$

The unit step function (Heaviside unit function) [7] $E(t)$ is defined as follows.

$$E(t) = \begin{cases} 0 & t < 0 \\ 1 & t \geq 0 \end{cases} \tag{B.23}$$

Consider evaluating the Fourier transform of $E(t)$ according to (5.31).

$$\mathcal{F}\{E(t)\} = \int_{-\infty}^{\infty} E(t)e^{-j\omega t}\,dt = \int_{0}^{\infty} e^{-j\omega t}\,dt$$

$$= \frac{1}{-j\omega}[e^{-j\omega t}]_0^{\infty} = \frac{1}{-j\omega}[\cos\omega t - j\sin\omega t]_0^{\infty} \tag{B.24}$$

On the last term of (B.24) both the real and imaginary parts oscillate when $t \to \infty$. Thus, the Fourier transform of the unit step function in the form of (B.24) is undetermined. So, we express the unit function as follows.

$$E_\alpha(t) = \begin{cases} 0 & t < 0 \\ e^{-\alpha t} & t \geq 0 \end{cases} \tag{B.25}$$

Here $\alpha > 0$ and we use the limit case $\alpha \to 0$. Figure B.6a compares the unit step functions expressed with (B.23) and (B.25) with $\alpha = 0.001$. Practically no difference is observed between the two plots.

With expression (B.25) and (5.31), the Fourier transform is evaluated as follows.

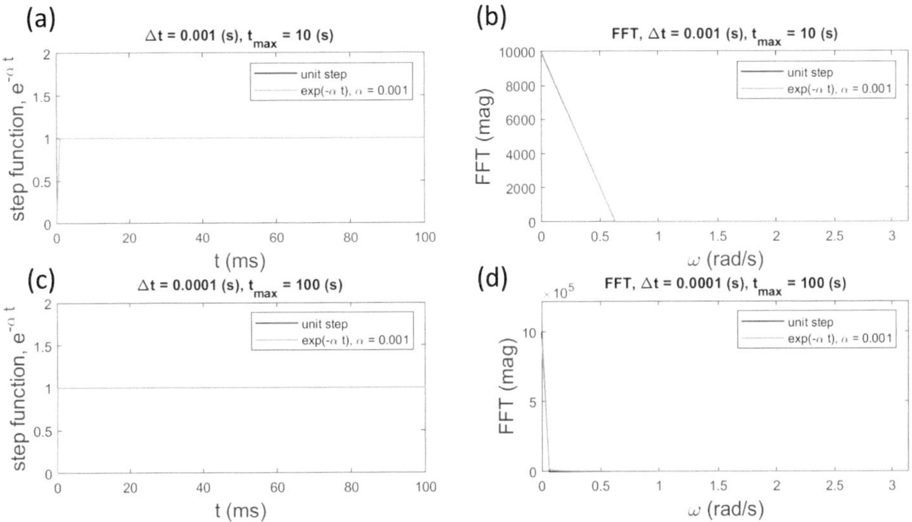

**Fig. B.6** Unit step functions and approximation with expression (B.24). **a** time domain signal with coarse time step and frequency resolution; **b** Fourier transform of **a**; **c** time domain signal with fine time step and frequency resolution; **d** Fourier transform of **c**

$$\mathcal{F}\{E_\alpha(t)\} = \int_{-\infty}^{\infty} e^{-\alpha t} e^{-j\omega t} dt = \int_{0}^{\infty} e^{-(\alpha+j\omega)t} dt$$

$$= \frac{1}{-(\alpha+j\omega)}\left[ e^{-\alpha t}\left( \cos\omega t - j\sin\omega t \right) \right]_0^{\infty}$$

$$= \frac{-1}{-(\alpha+j\omega)} = \frac{\alpha}{\alpha^2+\omega^2} - j\frac{\omega}{\alpha^2+\omega^2} \tag{B.26}$$

The real part takes the form of the above-mentioned Lorentzian before the normalization. Thus, in the $\alpha \to 0$ limit, the real part reduces to $\pi \lim_{\epsilon \to 0} L_n(\omega) = \pi\delta(\omega)$ (see (B.22)). The imaginary part of (B.26) is readily evaluated by substituting $\alpha = 0$.

$$-j\frac{\omega}{\alpha^2+\omega^2} \to -j\frac{1}{\omega} = \frac{1}{j\omega} \tag{B.27}$$

Thus, we find

$$\mathcal{F}\{E_\alpha(t)\} = F(\omega) = \pi\delta(\omega) + \frac{1}{j\omega} \tag{B.28}$$

The first term on the right-hand side is infinite at $\omega = 0$ and zero otherwise. The second term decreases inversely proportional to the frequency. This means that the Fourier transform of the unit step function is largely weighted on the zero frequency term (called the DC term) and decays with frequency.

Figure B.6 shows sample unit step functions along with the approximation with the exponential expression (B.24). Here Fig. B.6a, b are the time domain and frequency domain signal expressed with the time step $\delta t = 0.001$ s and the time duration of 10 s (the frequency resolution of 0.1 Hz). Figure B.6c, d correspond to (a) and (b) with 10 times finer time step and frequency resolution. The plot labeled "unit step" is expressed as the signal at the initial time bin is zero and the signal in all the other bins are unity. For the exponential expression (B.24), the value of the parameter $\epsilon$ is set to 0.001.

Figure B.6a, c indicate that at both time steps, the exponential expression (B.24) does not show any visible difference from the "unit step" plot. In the frequency domain signals, the finer frequency case shows a narrower delta function at the DC (0 Hz) end, as expected. In the coarse frequency resolution case in Fig. B.6b the Fourier spectra resulting from the "unit step" case and the exponential approximation case do not show a difference. In the fine frequency resolution case, the exponential approximation indicates a slight rise from the floor at the bottom of the delta function.

The above argument regarding the frequency dependence of the unit step function is easily understood intuitively as follows. Consider that the step function represents the motion of an object. The object is initially still at $t = 0$ and moves at the constant speed of unity for $t > 0$. If we observe this motion, it is apparent that the motion does not have an oscillatory behavior. This is represented by $\delta(\omega)$ in the first term on the right-hand side of (B.28) (DC, or 0 Hz).

It is interesting to note that the relation between the time domain and frequency domain signals is reciprocal between the impulse function and step function. Tapping in the time domain generates a flat frequency domain signal. Contrary, a flat time domain signal generates an impulse frequency domain signal. This relation is naively related to the following situations. The delta function is the temporal differentiation of a step function. The time domain differentiation corresponds to the multiplication of $j\omega$ ($j^2 = -1$ and $\omega$: angular frequency) in the frequency domain as will be discussed in the next section. The Fourier spectrum of the unit step function has $1/j\omega$ term. Hence, when multiplied by $j\omega$, it becomes unity. This is the Fourier spectrum of the delta function as the differential of the step function. The first term $\pi\delta(\omega)$ is interpreted as the initial value on the frequency axis. In the differentiation operation, this term vanishes as a constant.

## B.3  Rectangle Signal

Figure B.7 is the time and frequency domain plots for a rectangle signal for two signal widths. It is interesting to note that the shape of the Fourier spectrum is in between those of the impulse and step function cases (see Figs. B.1 and B.6d. Moreover, as the signal width decreases the spectrum flattens indicating the transition from the impulse spectrum for the step function case to the flat spectrum for the impulse case.

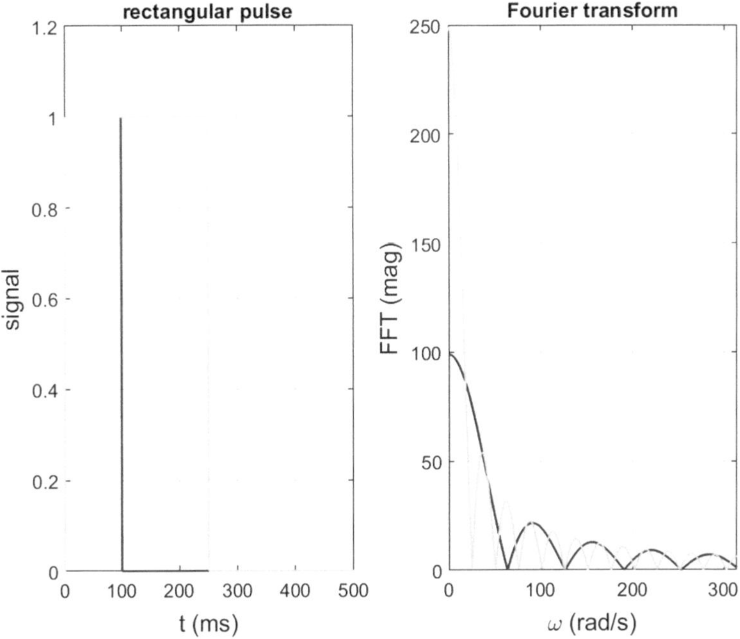

**Fig. B.7** Rectangle functions with two widths. **a** time domain signal; **b** Fourier transform of **a**

## B.4  Constant Signal

Define the following function to represent a constant of unity.

$$g(t) = 1 \quad -\infty < t < \infty \tag{B.29}$$

We can interpret $g(t)$ as the superposition of the unit step function defined for the negative and positive time axes; $E_-(t)$ and $E_+(t)$. Thus, the Fourier transform of $g(t)$ is the addition of the Fourier transform of these $E_-(t)$ and $E_+(t)$. From (B.24) and (B.28),

$$\mathcal{F}\{E(t)_+\} = \int_0^\infty e^{-j\omega t}dt = \pi\delta(\omega) + \frac{1}{j\omega} \tag{B.30}$$

Similarly,

$$\mathcal{F}\{E_-(t)\} = \int_{-\infty}^0 e^{-j\omega t}dt = \int_\infty^0 e^{-j\omega(-\tau)}(-d\tau) = \int_0^\infty e^{-j\omega(-\tau)}d\tau$$

$$= \int_0^\infty e^{-j(-\omega)\tau}d\tau = \pi\delta(-\omega) - \frac{1}{j\omega} \tag{B.31}$$

Since $\delta(-\omega) = \delta(\omega)$,

$$\mathcal{F}\{E_-(t) + E_+(t)\} = \left(\pi\delta(-\omega) - \frac{1}{j\omega}\right) + \left(\pi\delta(\omega) + \frac{1}{j\omega}\right) = 2\pi\delta(\omega) \qquad \text{(B.32)}$$

Thus we find the Fourier transform of $g(t) = 1$ as $2\pi\delta(\omega)$.

## B.5  Cosine and Sine Functions $\cos at, \sin at$

$$\mathcal{F}\{\cos at\} = \int_{-\infty}^{\infty} \left(\frac{e^{-jat} + e^{jat}}{2}\right) e^{-j\omega t} dt$$

$$= \int_{-\infty}^{\infty} \frac{1}{2} e^{-j(\omega+a)t} dt + \int_{-\infty}^{\infty} \frac{1}{2} e^{-j(\omega-a)t} dt \qquad \text{(B.33)}$$

We can interpret the first and second terms on the right-hand side of (B.33), respectively, as the Fourier transform of constant unity with the frequency shifts of $\omega \to \omega + a$ and $\omega \to \omega - a$. Thus, with the help of (B.32) we find the Fourier transform of this cosine function as follows.

$$\mathcal{F}\{\cos at\} = \frac{1}{2}2\pi\delta(\omega + a) + \frac{1}{2}2\pi\delta(\omega - a) = \pi\delta(\omega + a) + \pi\delta(\omega - a) \qquad \text{(B.34)}$$

Similarly,

$$\mathcal{F}\{\sin at\} = \int_{-\infty}^{\infty} \left(-\frac{e^{-jat} + e^{jat}}{2j}\right) e^{-j\omega t} dt$$

$$= \int_{-\infty}^{\infty} \frac{j}{2} e^{-j(\omega+a)t} dt - \int_{-\infty}^{\infty} \frac{j}{2} e^{-j(\omega-a)t} d$$

$$= j\{\pi\delta(\omega + a) - \pi\delta(\omega - a)\}$$

# Some Note on FFT C

Consider a sine function expressed with the same total time duration and two time-steps, 0.1 s (the coarse $\delta t$) and 0.02 s (the fine $\delta t$). The sine function has a frequency of 0.5 Hz and the time duration is 80 s. Figure C.1a presents the time series of this function for two cycles. Since the period of this function is 1/0.5 (Hz) $= 2$ s, the coarse one presents one period with 2 (s)/0.1 (s) $= 20$ data points. Similarly, the fine one presents one period with 100 data points.

The frequency resolution is determined as $1/40 = 0.025$ Hz for both cases. The maximum frequency (the sampling rate) of the coarse and fine cases are, respectively, 1/0.1 (s) $= 10$ Hz and 1/0.02 (s) $= 50$ Hz. Therefore, the total number of the frequency bins are 10 (Hz)/0.025 (Hz) $= 400$, and 50 (Hz)/0.025 (Hz) $= 2{,}000$. Figure C.1b presents the FFT magnitude as a function of the total frequency bin number for both cases.

Figure C.1c, d present the spectral peak for the coarse and fine time step cases. Notice that the peak frequency is at 0.5 Hz in both cases. Since the frequency resolution is commonly 0.025 Hz for both cases, 0.5 Hz corresponds to $0.5/0.025 = 20^{th}$ frequency bin for the coarse and fine time step cases. The difference regarding the frequency bin between the two cases is that the fine time step case has five times higher maximum frequency.

While the peak frequency is the same for the two cases, careful examination indicates that the foot of the spectral peak is somewhat raised from the floor as the two arrows in Fig. C.1c indicate. Theoretically, $\sin(\pi t)$ has a peak at 0.5 Hz with no width. When the maximum frequency is 50 Hz, the number of harmonics is not enough to represent the sharp rise at the spectral peak.

This simple example indicates that the FFT operation expresses the mathematical or physically correct peak frequency correctly regardless of the time step as long as it is small enough to present the time-domain signal accurately. The smaller the time step, therefore the higher the maximum frequency, the Fourier spectrum is expressed more accurately as well.


S. Yoshida, *Physics and Mathematics Behind Wave Dynamics*, Synthesis Lectures on Wave Phenomena in the Physical Sciences,
https://doi.org/10.1007/978-3-031-60354-9

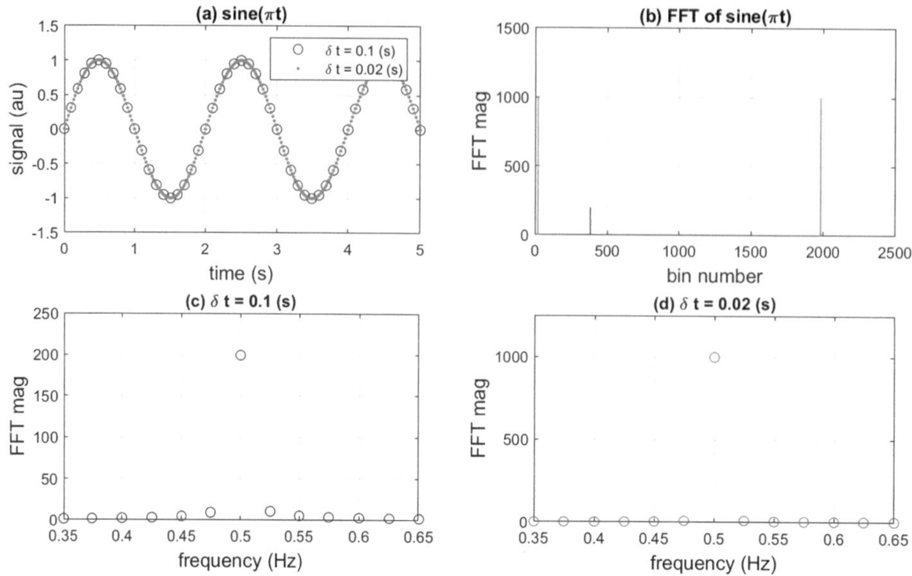

**Fig. C.1** Time and frequency series of a sine function. **a** Time series, **b** FFT magnitude versus bin, **c** FFT magnitude vs frequency (coarse time step), **d** FFT magnitude vs frequency (fine time step). The arrows in **c** indicate the raised foot due to insufficient of bin number

# References


1. Hoskins RF (2009) Delta functions: introduction to generalised functions, 2nd edn. Horwood Publishing, Ltd., Oxford, UK (2009)
2. Kuo HH (1996) White noise distribution theory, 1st edn. CRC Press
3. Stenger F (1993) Numerical methods based on sinc and analytic functions, 1st edn. Springer, New York, Berlin
4. Walck C (2007) Hand-book on statistical distributions for experimentalists, Internal Report SUF-PFY/96-01, University of Stockholm, last modified on September 10, 2007
5. Krishnomoorthy K (2019) Handbook of statistical distributions with applications, 2nd edn. CRC Press, Boca Raton, New York, p 343
6. Yariv A (1971) Introduction to optical electronics. Holt, Rinehart and Winston, New York, USA
7. Bird J (2017) The Laplace transform of the Heaviside function. In: Higher engineering mathematics, 8th edn. Routledge, London, Chap. 70

# Index




© The Editor(s) (if applicable) and The Author(s), under exclusive license to Springer
Nature Switzerland AG 2025
S. Yoshida, *Physics and Mathematics Behind Wave Dynamics*, Synthesis Lectures
on Wave Phenomena in the Physical Sciences,
https://doi.org/10.1007/978-3-031-60354-9

If you have any concerns about our products,
you can contact us on
ProductSafety@springernature.com

In case Publisher is established outside the EU,
the EU authorized representative is:
**Springer Nature Customer Service Center GmbH**
**Europaplatz 3, 69115 Heidelberg, Germany**

Printed by Libri Plureos GmbH
in Hamburg, Germany